WITHDRAWN FROM STOCK

Inheritance, Variation and Evolution

Ecosystems

ACKNOWLEDGEMENTS

The author and publisher are grateful to the copyright holders for permission to use quoted materials and images.

Images on page 42 from www.forestry.gov.uk
Image A © Forestry Commission
Image B Forestry Commission / Thomas Kirisits

Page 67 ©2009 Jupiterimages Corporation, ©Hemera/Thinkstock

Page 77 ©iStockphoto.com

Page 89 ©2009 Jupiterimages Corporation

All other images are ©Shutterstock.com
Illustrations © HarperCollinsPublishers

Every effort has been made to trace copyright holders and obtain their permission for the use of copyright material. The author and publisher will gladly receive information enabling them to rectify any error or omission in subsequent editions. All facts are correct at time of going to press.

Published by Letts Educational
An imprint of HarperCollinsPublishers
1 London Bridge Street
London SE1 9GF

ISBN: 9780008318352

Content first published 2016
This edition published 2019

10 9 8 7 6 5 4 3 2 1

© HarperCollinsPublishers Limited 2019

All rights reserved. No part of this publication may be reproduced, stored in a retrieval system, or transmitted, in any form or by any means, electronic, mechanical, photocopying, recording or otherwise, without the prior permission of Letts Educational.

British Library Cataloguing in Publication Data.

A CIP record of this book is available from the British Library.

Series Concept: Emily Linnett
Author: Tom Adams
Commissioning and Series Editor: Charlotte Christensen
Editor and Project Manager: Tracey Cowell
Inside Concept Design: Paul Oates
Cover Design: Sarah Duxbury
Text Design, Layout and Artwork: Q2A Media
Production: Karen Nulty
Printed by RRD South China

Structure of cells

Cells are the basis of life. All processes of life take place within them. There are many different kinds of specialised cell but they all have several common features. The two main categories of cells are **prokaryotic** (prokaryotes) and **eukaryotic** (eukaryotes).

Prokaryotes

Prokaryotes are simple cells such as bacteria and archaebacteria. Archaebacteria are ancient bacteria with different types of cell wall and genetic codes to other bacteria. They include microorganisms that can use methane or sulfur as a means of nutrition (methanogens and thermoacidophiles).

Prokaryotic cells:

➤ are smaller than eukaryotic cells
➤ have cytoplasm, a cell membrane and a cell wall
➤ have genetic material in the form of a DNA loop, together with rings of DNA called plasmids
➤ do not have a nucleus.

The chart below shows the relative sizes of prokaryotic and eukaryotic cells, together with the magnification range of different viewing devices.

A bacterium

Plasmid DNA: a small, commonly circular, section of DNA that can replicate independently of chromosomal DNA

Ribosomes: sites of protein manufacture

Chromosomal DNA: the DNA of bacteria is not found within a nucleus and is usually found as one circular chromosome

Cytoplasm

Cell wall: provides structural support to the bacteria (is not made of cellulose)

Flagella: tail-like structures that rotate to help some bacteria move

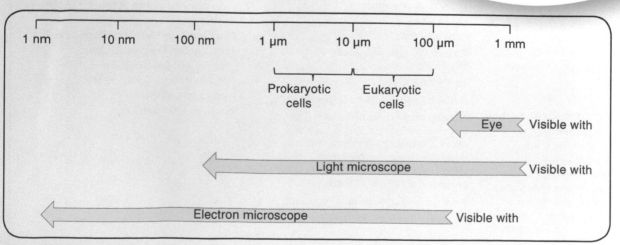

| 1 nm | 10 nm | 100 nm | 1 µm | 10 µm | 100 µm | 1 mm |

Prokaryotic cells
Eukaryotic cells

Eye — Visible with

Light microscope — Visible with

Electron microscope — Visible with

Eukaryotes

Eukaryotic cells:

➤ are more complex than prokaryotic cells (they have a cell membrane, cytoplasm and genetic material enclosed in a nucleus)

➤ are found in animals, plants, fungi (e.g. toadstools, yeasts, moulds) and protists (e.g. amoeba)

➤ contain membrane-bound structures called **organelles**, where specific functions are carried out.

Here are the main organelles in plant and animal cells.

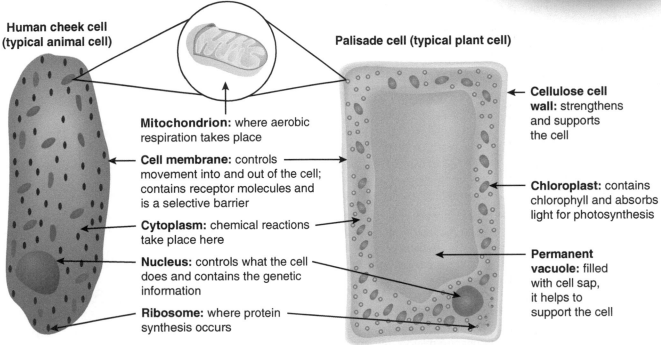

Human cheek cell (typical animal cell)

Palisade cell (typical plant cell)

Mitochondrion: where aerobic respiration takes place

Cell membrane: controls movement into and out of the cell; contains receptor molecules and is a selective barrier

Cytoplasm: chemical reactions take place here

Nucleus: controls what the cell does and contains the genetic information

Ribosome: where protein synthesis occurs

Cellulose cell wall: strengthens and supports the cell

Chloroplast: contains chlorophyll and absorbs light for photosynthesis

Permanent vacuole: filled with cell sap, it helps to support the cell

Plant cells tend to be more regular in shape than animal cells. They have additional structures: cell wall, sap vacuole and sometimes chloroplasts.

Create a poster illustrating the main features of bacterial, plant and animal cells (i.e. prokaryotes and eukaryotes).

➤ Use different colours for the main organelles, e.g. blue for the nucleus, orange for cytoplasm.

➤ Add labels that describe the functions of the organelles.

Keyword

Organelle ➤ A membrane-bound structure within a cell that carries out a particular function

1. List three examples of prokaryotes and three examples of eukaryotes.
2. How are prokaryotes and eukaryotes different in terms of genetic information?
3. All cells have a cell membrane. True or false?
4. Which organelle carries out the function of respiration?
5. Algae are a type of plant. Why do the cells contain chloroplasts?

Organisation and differentiation

Multicellular organisms need to have a coordinated system of structures so they can carry out vital processes, e.g. respiration, excretion, nutrition, etc.

Principles of organisatio

Cells are the most basic unit of living organisms.

Tissues are collections of simil cells that work together to carry o the same functior

An organ is formed when a group of tissues combine together and do a particular job.

Organs form organ systems, e.g. the intestines and the stomach are part of the digestive system.

Organ systems work together to make up a complex multicellular organism.

Organisation

Cell specialisation

Animals and plants have many different types of cells. Each cell is adapted to carry out a specific function. Some cells can act independently, e.g. white blood cells, but most operate together as tissues.

Type of specialised animal cell	How they are adapted
Sperm cells	➤ They are adapted for swimming in the female reproductive system – mitochondria in the neck release energy for swimming. ➤ They are adapted for carrying out fertilisation with an egg cell – the acrosome contains enzymes for digestion of the ovum's outer protective cells at fertilisation. Mitochondria, End piece, Haploid nucleus, Acrosome
Egg cells (ovum)	➤ They are very large in order to carry food reserves for the developing embryo. ➤ After fertilisation, the cell membrane changes and locks out other sperm. Cell membrane, Cytoplasm, Mitochondria, Haploid nucleus
Ciliated epithelial cells	➤ They line the respiratory passages and help protect the lungs against dust and microorganisms. Mucus, Gland cells, Cilia, Movement of mucus
Nerve cells	➤ They have long, slender extensions called axons that carry nerve impulses.
Muscle cells	➤ They are able to contract (shorten) to bring about the movement of limbs.

Type of specialised plant cell	How they are adapted
Root hair cells	➤ They have tiny, hair-like extensions. These increase the surface area of roots to help with the absorption of water and minerals.
Xylem	➤ They are long, thin, hollow cells. Their shape helps with the transport of water through the stem, roots and leaves.
Phloem	➤ They are long, thin cells with pores in the end walls. Their structure helps the cell sap move from one phloem cell to the next.

Cell differentiation and stem cells

Stem cells are found in animals and plants. They are unspecialised or **undifferentiated**, which means that they have the potential to become almost any kind of cell. Once the cell is fully specialised, it will possess sub-cellular organelles specific to the function of that cell. Embryonic stem cells, compared with those found in adults' bone marrow, are more flexible in terms of what they can become.

Animal cells are mainly restricted to repair and replacement in later life. Plants retain their ability to **differentiate** (specialise) throughout life.

Using stem cells in humans

Therapeutic cloning treats conditions such as diabetes. Embryonic stem cells (that can specialise into any type of cell) are produced with the same genes as the patient. If these are introduced into the body, they are not usually rejected.

Treating paralysis is possible using stem cells that are capable of differentiating into new nerve cells. The brain uses these new cells to transmit nervous impulses to the patient's muscles.

There are benefits and objections to using stem cells.

Benefits	Risks and objections
➤ Stem cells left over from in vitro fertilisation (IVF) treatment (that would otherwise be destroyed) can be used to treat serious conditions. ➤ Stem cells are useful in studying how cell division goes wrong, e.g. cancer. ➤ In the future, stem cells could be used to grow new organs for transplantation.	➤ Some people believe that an embryo at any age is a human being and so should not be used to grow cells or be experimented on. ➤ One potential risk of using stem cells is transferring viral infections. ➤ If stem cells are used in an operation, they might act as a reservoir of cancer cells that spread to other parts of the body.

Using stem cells in plants

Plants have regions of rapid cell division called **meristems**. These growth regions contain stem cells that can be used to produce clones cheaply and quickly.

Meristems can be used for:
➤ growing and preserving rare varieties to protect them from extinction
➤ producing large numbers of disease-resistant crop plants.

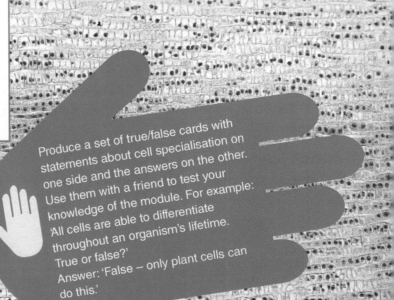

Produce a set of true/false cards with statements about cell specialisation on one side and the answers on the other. Use them with a friend to test your knowledge of the module. For example: 'All cells are able to differentiate throughout an organism's lifetime. True or false?' Answer: 'False – only plant cells can do this.'

1. In the body, what do you call an arrangement of different tissues that carry out a particular job?
2. Describe how a muscle cell is specialised to perform its function.
3. Which type of specialised plant cell is used to transport the sugars made in photosynthesis?
4. Give one benefit of and one objection to the scientific use of stem cells.

Microscopy and microorganisms

Microscopes

Microscopes:
➤ observe objects that are too small to see with the naked eye
➤ are useful for showing detail at cellular and sub-cellular level.

There are two main types of microscope: the **light microscope** and the **electron microscope**.

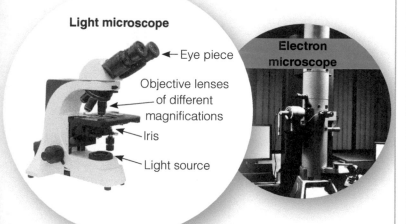

Light microscope
← Eye piece
Objective lenses of different magnifications
Iris
Light source

Electron microscope

You will probably use a light microscope in your school laboratory.

The electron microscope was invented in 1931. It has increased our understanding of sub-cellular structures because it has much higher magnifications and resolution than a light microscope.

These are white blood cells, as seen through a transmission electron microscope. The black structures are nuclei.

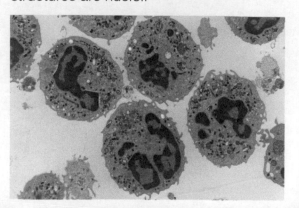

Here are some common units used in microscopy.

Measure	Scale	Symbol
1 metre		m
1 millimetre	$\frac{1}{1000}$ (a thousandth) of a metre ($\times 10^{-3}$)	mm
1 micrometre	$\frac{1}{1\,000\,000}$ (a millionth) of a metre ($\times 10^{-6}$)	μm
1 nanometre	$\frac{1}{1\,000\,000\,000}$ (a thousand millionth or a billionth) of a metre ($\times 10^{-9}$)	nm
1 picometre	$\frac{1}{1\,000\,000\,000\,000}$ (a million millionth or a trillionth) of a metre ($\times 10^{-12}$)	pm

Comparing light and electron microscopes

Light microscope	Electron microscope
Uses light waves to produce images	Uses electrons to produce images
Low resolution	High resolution
Magnification up to ×1500	Magnification up to ×500 000 (2D) and ×100 000 (3D)
Able to observe cells and larger organelles	Able to observe small organelles
2D images only	2D and 3D images produced

Keywords

Resolution ➤ The smallest distance between two points on a specimen that can still be told apart
Pathogen ➤ A harmful microorganism
Binary fission ➤ Process where bacterial cells divide into two
Colony ➤ A large number of microorganisms (of one type) growing in a location, e.g. a circular colony on the surface of agar

Magnification and resolution

Magnification measures how many times an object under a microscope has been made larger.

To calculate the magnifying power of a microscope, use this formula:

$$\text{magnification} = \frac{\text{size of image}}{\text{size of real object}}$$

You may be asked to carry out calculations using magnification.

> **Example:** A light microscope produces an image of a cell which has a diameter of 1500 μm. The cell's actual diameter is 50 μm. Calculate the magnifying power of the microscope.
> $$\text{magnification} = \frac{1500}{50} = \times 30$$

Resolution is the smallest distance between two points on a specimen that can still be told apart. The diagram below shows the limits of resolution for a light microscope. When looking through it you can see fine detail down to 200 nanometres. Electron microscopes can see detail down to 0.05 nanometres!

 200 nm

Staining techniques are used in light and electron microscopy to make organelles more visible.
- **Methylene blue** is used to stain the nuclei of animal cells for viewing under the light microscope.
- **Heavy metals** such as cadmium can be used to stain specimens for viewing under the electron microscope.

Culturing microorganisms

To view microorganisms under the microscope, pure, uncontaminated samples are grown or **cultured** in the laboratory. These cultures can be used for research, e.g. investigating the action of disinfectants or antibiotics.

In order to grow, microorganisms need:
- **food** in the form of a growth medium; in the laboratory, this is usually nutrient agar or broth
- **warmth** – a temperature of 25°C is used in school laboratories, as it lowers the possibility of **pathogens** growing
- **moisture**.

Aseptic technique

Aseptic technique is needed to culture microorganisms in the laboratory so that:
- uncontaminated cultures are produced
- the risk of scientists being infected by pathogens is reduced.

Bacterial growth

Under ideal conditions, a bacterial culture will grow quickly over a period of days by **binary fission**. In this process, a cell divides into two in as little time as twenty minutes!

The bacteria continue to divide in this way: 1, 2, 4, 8, 16, 32, 64, etc. So after a day, the numbers are in their millions. As they grow, the individual cells merge to produce circular **colonies** that are visible to the naked eye. You can count them or calculate their area to estimate the growth rate.

> **WS** **Calculating the cross-sectional area of a bacterial colony**
>
> Bacterial colonies grow **radially**, i.e. in circles. To calculate a rate of growth, the area of the colony can be worked out at two points in time.
>
> **Example:** A colony of *M. luteus* is observed growing on agar in a petri dish. The diameter is measured with a ruler and is 20 mm. Exactly 24 hours later, the colony's diameter is 30 mm. What is its growth rate?
> - Area of colony for first measurement = πr^2 where $\pi = 3.14$ and $r = \frac{1}{2} \times 20$
>   Area = $3.14 \times 10^2 = 314\,\text{mm}^2$
> - Area of colony for second measurement = $3.14 \times 15^2 = 706.5\,\text{mm}^2$
> - Rate of growth = $\dfrac{706.5 - 314}{24}$
> $$= \mathbf{16.35\,mm^2\,h^{-1}}$$

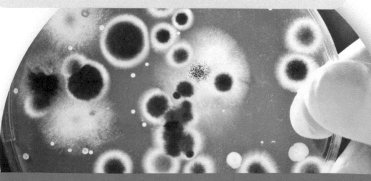

1. List two differences between light and electron microscopes.
2. Which three conditions are needed for microorganisms to grow?

Cell division

Multicellular organisms grow and reproduce using cell division and cell enlargement. Plants can also grow via differentiation into leaves, branches, etc.

Cell division begins from the moment of fertilisation when the zygote replicates itself exactly through **mitosis**. Later in life, an organism may use cell division to produce sex cells (**gametes**) in a different type of division called **meiosis**.

The different stages of cell division make up the **cell cycle** of an organism.

Keywords

Gamete ➤ A sex cell, i.e. sperm or egg

Daughter cells ➤ Multiple cells arising from mitosis and meiosis

Polymer ➤ Large molecule, made of repeating units called monomers

Diploid ➤ A full set of chromosomes in a cell (twice the haploid number)

Haploid ➤ A half set of chromosomes in a cell; haploid cells are either eggs or sperm

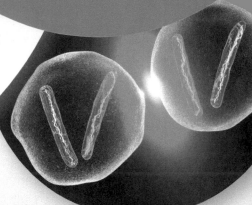

Chromosomes

Chromosomes are found in the nucleus of eukaryotic cells. They are made of DNA and carry a large number of genes. Chromosomes exist as pairs called **homologues**.

For cells to duplicate exactly, it is important that all of the genetic material is duplicated. Chromosomes take part in a sequence of events that ensures the genetic code is transmitted precisely and appears in the new **daughter cells**.

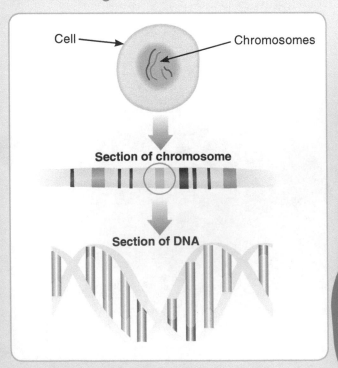

Cell — Chromosomes

Section of chromosome

Section of DNA

DNA replication

During the cell cycle, the genetic material (made of the **polymer** molecule, DNA) is doubled and then divided between the identical daughter cells. This process is called **DNA replication**. (See Module 29 for more information about DNA.)

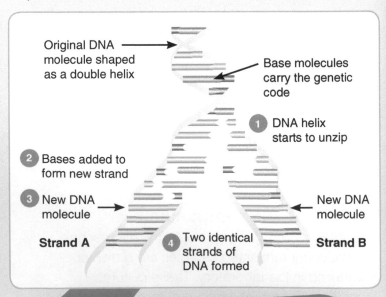

Original DNA molecule shaped as a double helix

Base molecules carry the genetic code

1 DNA helix starts to unzip

2 Bases added to form new strand

3 New DNA molecule

New DNA molecule

Strand A

4 Two identical strands of DNA formed

Strand B

Use plasticine or playdough to create models showing mitosis and meiosis. Mitosis should show four stages and meiosis six stages. You could:
➤ use different colours for the chromosomes and the cytoplasm
➤ try mixing the stages up and then re-arranging them from memory
➤ explain to a friend what is happening at each stage in terms of the chromosomes and cytoplasm.

Mitosis

Mitosis is where a **diploid** cell (one that has a complete set of chromosomes) divides to produce two more diploid cells that are genetically identical. Most cells in the body are diploid.

Humans have a diploid number of 46.

Mitosis produces new cells:

➤ for growth
➤ to replace old cells
➤ to repair damaged tissue
➤ for asexual reproduction.

Before the cell divides, the DNA is duplicated and other organelles replicate, e.g. mitochondria and ribosomes. This ensures that there is an exact copy of all the cell's content.

Mitosis – the cell copies itself to produce two genetically identical cells

Parent cell with two pairs of chromosomes.

Each chromosome replicates (copies) itself.

Each 'daughter' cell has the same number of chromosomes, and contains the same genes, as the parent cell.

Chromosomes line up along the centre of the cell, separate into chromatids and move to opposite poles.

Meiosis

Meiosis takes place in the testes and ovaries of sexually reproducing organisms and produces gametes (eggs or sperm). The gametes are called **haploid** cells because they contain half the number of chromosomes as a diploid cell. This chromosome number is restored during **fertilisation**.

Humans have a haploid number of 23.

Meiosis – the cell divides twice to produce four cells with genetically different sets of chromosomes

Cell with two pairs of chromosomes (diploid cell).

Each chromosome replicates itself.

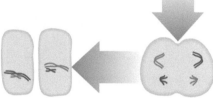

Cell divides for the first time.

Chromosomes part company and move to opposite poles.

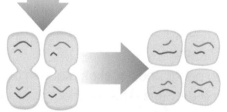

Copies now separate and the second cell division takes place.

Four haploid cells (gametes), each with half the number of chromosomes of the parent cell.

Cancer

Cancer is a non-infectious disease caused by **mutations** in living cells.

Cancerous cells:

➤ divide in an uncontrolled way
➤ form **tumours**.

Benign tumours do not spread from the original site of cancer in the body. **Malignant tumour cells** invade neighbouring tissues. They spread to other parts of the body and form **secondary tumours**.

Making healthy lifestyle choices is one way to reduce the likelihood of cancer. These include:

➤ not smoking tobacco products (cigarettes, cigars, etc.)
➤ not drinking too much alcohol (causes cancer of the liver, gut and mouth)
➤ avoiding exposure to UV rays (e.g. sunbathing, tanning salons)
➤ eating a healthy diet (high fibre reduces the risk of bowel cancer) and doing moderate exercise to reduce the risk of obesity.

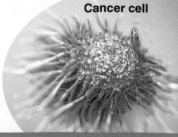

Cancer cell

1. Why do multicellular organisms carry out cell division?
2. Name the structures in the nucleus that carry genetic information.
3. What are the purposes of mitosis and meiosis?
4. Which structures are replicated during cell division?

Metabolism – respiration

Keywords

Exothermic reaction ➤ A reaction that gives out heat

Energy demand ➤ Energy required by tissues (particularly muscle) to carry out their functions

Oxygen debt ➤ The oxygen needed to remove lactic acid after exercise

Metabolism

Metabolism is the sum of all the chemical reactions that take place in the body.

The two types of metabolic reaction are:
➤ building reactions (**anabolic**)
➤ breaking-down reactions (**catabolic**).

Anabolic reactions

Anabolic reactions require the input of energy. Examples include:
➤ converting glucose to starch in plants, or glucose to glycogen in animals
➤ the synthesis of lipid molecules

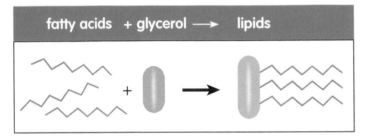

fatty acids + glycerol ⟶ lipids

➤ the formation of **amino acids** in plants (from glucose and nitrate ions) which, in turn, are built up into proteins.

Catabolic reactions

Catabolic reactions release energy. Examples include:
➤ breaking down amino acids to form **urea**, which is then excreted
➤ respiration.

Catabolic reactions produce waste energy in the form of heat (an **exothermic reaction**), which is transferred to the environment.

Respiration

Respiration continuously takes place in all organisms – the need to release energy is an essential life process. The reaction gives out energy and is therefore **exothermic**.

Aerobic respiration

Aerobic respiration takes place in cells. Oxygen and glucose molecules react and release energy. This energy is stored in a molecule called **ATP**.

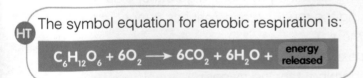

glucose + oxygen ⟶ carbon dioxide + water

(HT) The symbol equation for aerobic respiration is:

$$C_6H_{12}O_6 + 6O_2 \longrightarrow 6CO_2 + 6H_2O + \text{energy released}$$

Energy is used in the body for many processes, including:
➤ muscle contraction (for movement)
➤ active transport
➤ transmitting nerve impulses
➤ synthesising new molecules
➤ maintaining a constant body temperature.

Anaerobic respiration

Anaerobic respiration takes place in the absence of oxygen and is common in muscle cells. It quickly releases a **smaller** amount of energy than aerobic respiration through the **incomplete breakdown** of glucose.

glucose ⟶ lactic acid + energy released

In plant and yeast cells, anaerobic respiration produces different products.

glucose ⟶ ethanol + carbon dioxide + energy released

(HT) The symbol equation for anaerobic respiration in plant and yeast cells is:

$$C_6H_{12}O_6 \longrightarrow 2C_2H_5OH + 2CO_2 + \text{energy released}$$

This reaction is used extensively in the brewing and wine-making industries. It is also the initial process in the manufacture of spirits in a distillery.

Response to exercise

In animals, anaerobic respiration takes place when muscles are working so hard that the lungs and circulatory system cannot deliver enough oxygen to break down all the available glucose through aerobic respiration. In these circumstances the **energy demand** of the muscles is high.

Anaerobic respiration and recovery

Anaerobic respiration releases energy much faster over short periods of time. It is useful when short, intense bursts of energy are required, e.g. a 100 m sprint.

However, the incomplete oxidation of glucose causes **lactic acid** to build up. Lactic acid is toxic and can cause pain, cramp and a sensation of fatigue.

The lactic acid must be broken down quickly and removed to avoid cell damage and prolonged muscle fatigue.

➤ During exercise the body's heart rate, breathing rate and breath volume increase so that sufficient oxygen and glucose is supplied to the muscles, and so that lactic acid can be removed.

➤ This continues after exercise when deep breathing or panting occurs until all the lactic acid is removed. This repayment of oxygen is called **oxygen debt**.

HT
➤ Lactic acid is transported to the liver where it is converted back to glucose.
➤ Oxygen debt is the amount of **extra** oxygen that the body needs after exercise to react with the lactic acid and remove it from the cells.

WS You may be asked to present data as graphs, tables, bar charts or histograms. For example, you could show data of breathing and heart rates on a line graph.

What does the line graph below tell you about breathing and pulse rates during recovery? Why is a line graph a good way to present the data?

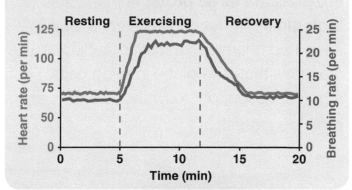

1. Give one example of a building reaction and one of a breaking-down reaction.
2. Which type of respiration releases most energy – aerobic or anaerobic?
3. Give two uses of energy release in the body.
4. Name the products of anaerobic respiration in humans and in yeast.
5. Why do athletes pant after a race?

5

Metabolism – enzymes

Enzymes are **large proteins** that act as **biological** catalysts. This means they speed up chemical reactions, including reactions that take place in living cells, e.g. respiration, photosynthesis and protein synthesis.

Enzyme facts

Enzymes:

➤ are specific, i.e. one enzyme catalyses one reaction

➤ have an active site, which is formed by the precise folding of the enzyme molecule

➤ can be denatured by high temperatures and extreme changes in pH

➤ have an **optimum temperature** at which they work – for many enzymes this is approximately 37°C (body temperature)

➤ have an optimum pH at which they work – this varies with the site of enzyme activity, e.g. pepsin works in the stomach and has an optimum pH of 1.5 (acid), salivary amylase works best at pH 7.3 (alkaline).

Keywords

Catalyst ➤ A substance that controls the rate of a chemical reaction without being chemically changed itself

Active site ➤ The place on an enzyme molecule into which a substrate molecule fits

Denaturation ➤ When a protein molecule such as an enzyme changes shape and makes it unable to function

Substrate ➤ The molecule acted on by an enzyme

Kinetic energy ➤ Energy possessed by moving objects, e.g. reactant molecules such as enzyme and substrate molecules

Enzyme activity

Enzyme molecules work by colliding with **substrate** molecules and forcing them to break up or to join with others in synthesis reactions. The theory of how this works is called the **lock and key theory**.

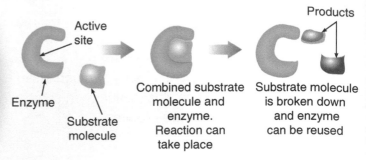

Active site

Enzyme

Substrate molecule

Combined substrate molecule and enzyme. Reaction can take place

Products

Substrate molecule is broken down and enzyme can be reused

High temperatures denature enzymes because excessive heat vibrates the atoms in the protein molecule, putting a strain on the bonds and breaking them. This changes the shape of the active site.

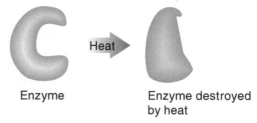

Heat

Enzyme

Enzyme destroyed by heat

In a similar way, an extreme pH alters the active site's shape and prevents it from functioning.

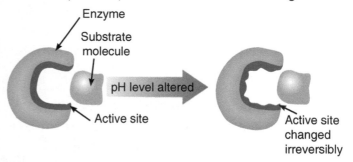

Enzyme

Substrate molecule

pH level altered

Active site

Active site changed irreversibly

At lower than the optimal temperature, an enzyme still works but much more slowly. This is because the low **kinetic energy** of the substrate and enzyme molecules lowers the number of collisions that take place. When they do collide, the energy is not always sufficient to create a bond between them.

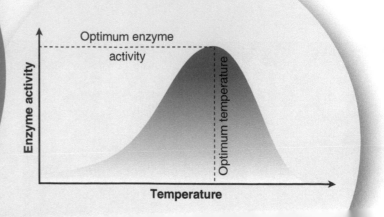

Optimum enzyme activity

Enzyme activity

Optimum temperature

Temperature

Enzymes in the digestive system

Enzymes in the digestive system help break down large nutrient molecules into smaller ones so they can be absorbed into the blood across the wall of the small intestine.

Enzyme types in the digestive system include carbohydrases, proteases and lipases.

Carbohydrases break down carbohydrates, e.g. **amylase**, which is produced in the mouth and small intestine.

starch ⟶ maltose

Other carbohydrases break down complex sugars into smaller sugars.

Proteases break down protein, e.g. **pepsin**, which is produced in the stomach.

protein ⟶ peptides ⟶ amino acids

Other enzymes in the small intestine complete protein breakdown with the production of amino acids.

Lipases, which are produced in the small intestine, break down lipids.

lipid ⟶ fatty acids + glycerol

Bile

Bile is a digestive chemical. It is produced in the liver and stored in the gall bladder.
➤ Its alkaline pH neutralises hydrochloric acid that has been produced in the stomach.
➤ It **emulsifies** fat, breaking it into small droplets with a large surface area.
➤ Its action enables lipase to break down fat more efficiently.

What happens to digested food?

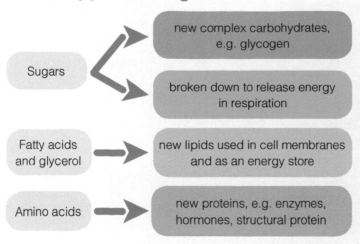

Sugars
- new complex carbohydrates, e.g. glycogen
- broken down to release energy in respiration

Fatty acids and glycerol
- new lipids used in cell membranes and as an energy store

Amino acids
- new proteins, e.g. enzymes, hormones, structural protein

Calorimetry

Enzyme-controlled metabolic reactions such as respiration release energy. The amount of energy contained in a substrate can be measured using **calorimetry**.

The substrate (food) is placed in a calorimeter and burned in pure oxygen. The energy released heats up the surrounding water and the temperature rise is measured.

The energy released per gram can be calculated.

$$\text{energy (J)} = \text{mass of water heated (g)} \times 4.2 \times \text{temperature rise (}^{\circ}\text{C)}$$

 During your course you will investigate how certain factors affect the rate of enzyme activity. These include temperature, pH and substrate concentration.

Design an investigation to discover how temperature affects the activity of amylase. Here are some guidelines.
➤ Iodine solution turns from a red-brown colour to blue-black in the presence of starch.
➤ You can measure amylase activity by timing how long it takes for iodine solution to stop turning blue-black.
➤ A water bath can be set up with a thermometer. Add cold or hot water to regulate the temperature.
➤ Identify the independent, dependent and control variables in the investigation. Write out your method in clear steps.

1. Name two factors that affect the rate of enzyme activity.
2. Describe, in simple terms, how lock and key theory explains enzyme action.
3. State two ways in which bile aids in the digestion of lipids.
4. Which molecules act as building blocks for protein polymers?

Mind map

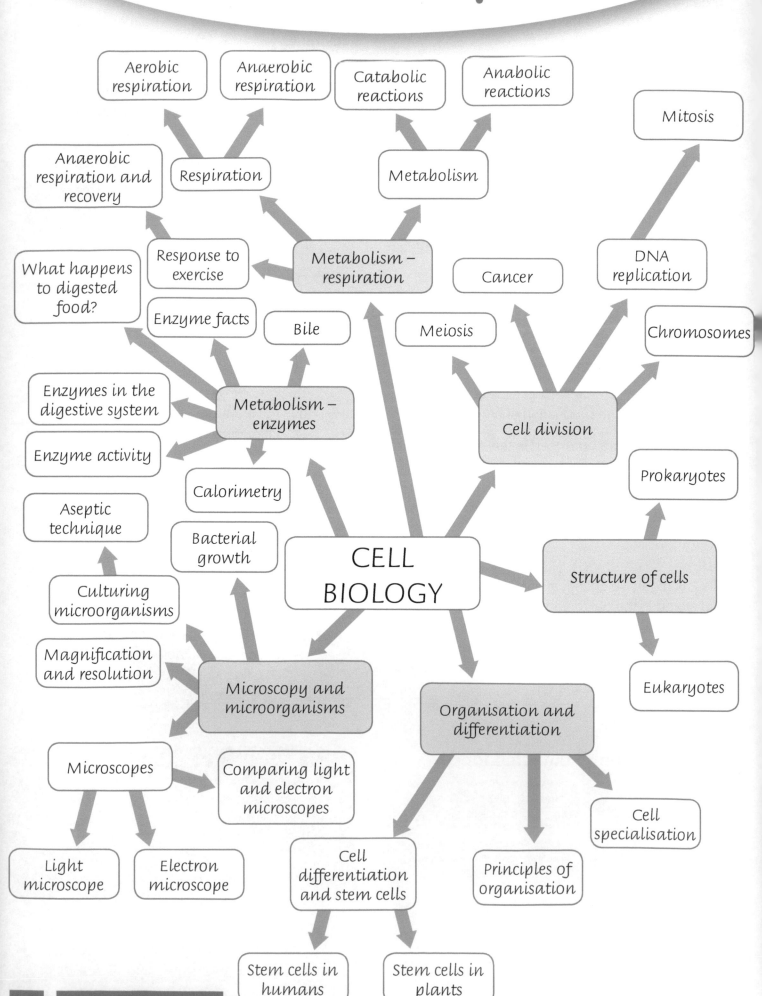

Aerobic respiration

Anaerobic respiration

Catabolic reactions

Anabolic reactions

Mitosis

Anaerobic respiration and recovery

Respiration

Metabolism

What happens to digested food?

Response to exercise

Metabolism – respiration

Cancer

DNA replication

Enzyme facts

Bile

Meiosis

Chromosomes

Enzymes in the digestive system

Metabolism – enzymes

Cell division

Enzyme activity

Prokaryotes

Aseptic technique

Calorimetry

CELL BIOLOGY

Structure of cells

Bacterial growth

Culturing microorganisms

Magnification and resolution

Microscopy and microorganisms

Organisation and differentiation

Eukaryotes

Microscopes

Comparing light and electron microscopes

Cell specialisation

Light microscope

Electron microscope

Cell differentiation and stem cells

Principles of organisation

Stem cells in humans

Stem cells in plants

Practice questions

1. Name the parts of the bacterial cell (**A–D**). **(4 marks)**

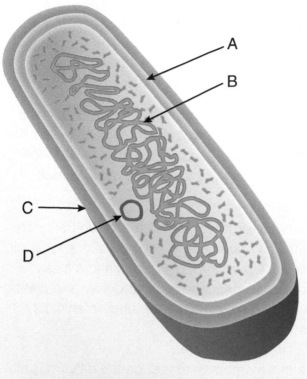

2. Lisa is playing football. She sprints the length of the pitch and scores a goal. However, she can barely celebrate because her legs have gone weak and she is panting heavily.

   a) Explain why Lisa's legs feel weak. **(2 marks)**

   b) Why would Lisa be unable to play the entire game at such a fast pace? **(1 mark)**

   c) Why does aerobic respiration yield more energy from glucose? **(1 mark)**

3. In 1894, the lock and key theory of enzyme action was put forward by Emil Fischer. The diagram shows the first stage in the reaction between an enzyme and a reactant.

   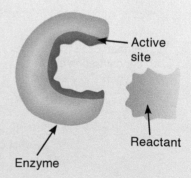

   a) What term describes the way that enzyme molecules change at high temperatures? **(1 mark)**

   b) Explain how a drop in pH will affect the structure of the enzyme shown in the diagram. **(2 marks)**

4. Barney has grown some bacterial colonies on an agar plate. He measures the radius of eight colonies. His results are shown in the table below.

   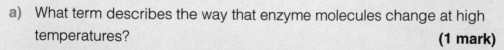

Colony	1	2	3	4	5	6	7	8
Radius (mm)	10	8	2	3	2	7	5	4

   Using the formula πr^2, calculate the **mean** cross-sectional area of the colonies. **(2 marks)**

Cell transport

Diffusion

Living cells need to obtain oxygen, glucose, water, mineral ions and other dissolved substances from their surroundings. They also need to excrete waste products, such as carbon dioxide or urea. These substances pass through the cell membrane by **diffusion**.

Diffusion:

➤ is the (net) movement of particles in a liquid or gas from a region of high concentration to one of low concentration (down a **concentration gradient**)

➤ happens due to the random motion of particles past each other

➤ stops once the particles have completely spread out

➤ is passive, i.e. requires no input of energy

➤ can be increased in terms of rate by making the concentration gradient steeper, the diffusion path shorter, increasing the temperature or increasing the surface area over which the process occurs, e.g. having a folded cell membrane.

A protist called amoeba can absorb oxygen through diffusion.

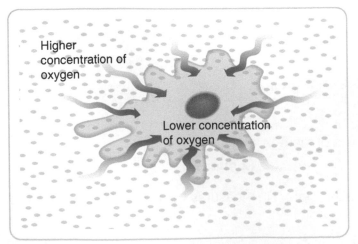

Higher concentration of oxygen

Lower concentration of oxygen

You may be asked to calculate rates of diffusion using **Fick's law**.

$$\text{rate of diffusion} \propto \frac{\text{surface area} \times \text{concentration difference}}{\text{thickness of membrane}}$$

$\propto$ means proportional to

Surface area to volume ratios

A **unicellular organism** (such as a protist) can absorb materials by diffusion directly from the environment. This is because it has a **large surface area to volume ratio**.

However, for a large, **multicellular organism**, the diffusion path between the environment and the inner cells of the body is long. Its large size also means that the **surface area to volume ratio** is **small**.

Adaptations

Multicellular organisms therefore need transport systems and specialised structures for exchanging materials, e.g. mammalian lungs and a small intestine, fish gills, and roots and leaves in plants. These increase diffusion efficiency in animals because they have:

➤ a large surface area

➤ a thin membrane to reduce the diffusion path

➤ an extensive blood supply for transport (animals)

➤ a ventilation system for gaseous exchange, e.g. breathing in animals.

In mammals, the individual air sacs in the lungs increase their surface area by a factor of thousands. Ventilation moves air in and out of the alveoli and the heart moves blood through the capillaries. This maintains the diffusion gradient. The capillary and alveolar linings are very thin, decreasing the diffusion path.

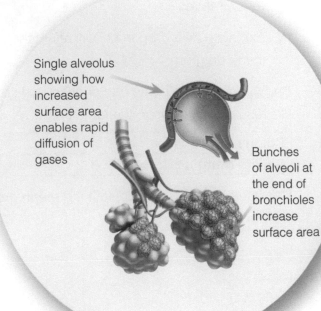

Single alveolus showing how increased surface area enables rapid diffusion of gases

Bunches of alveoli at the end of bronchioles increase surface area

Osmosis

Osmosis is a special case of diffusion that involves the movement of water only.
There are two ways of describing osmosis.

1. The net movement of **water** from a region of **low** solute concentration to one of **high** concentration.
2. The movement down a **water potential gradient**.

Osmosis:

➤ occurs across a **partially permeable membrane**, so solute molecules cannot pass through (only water molecules can)
➤ occurs in all organisms
➤ is passive (in other words, requires no input of energy)
➤ allows water movement into root hair cells from the soil and between cells inside the plant
➤ can be demonstrated and measured in plant cells using a variety of tissues, e.g. potato chips, daffodil stems.

Osmosis

| Dilute solution (high concentration of water) | Concentrated solution (low concentration of water) |

Partially permeable membrane

Net movement of water molecules

Active transport

Substances are sometimes absorbed **against** a concentration gradient, i.e. from a low to a high concentration.

Active transport:

➤ requires the release of energy from respiration
➤ takes place in the small intestine in humans, where sugar is absorbed into the bloodstream
➤ occurs via protein carrier molecules in the cell membrane
➤ allows plants to absorb mineral ions from the soil through root hair cells.

A cell absorbing ions by active transport

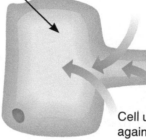

Root hair cell with high concentration of nitrate ions

Soil with lower concentration of nitrate ions

Cell uses energy to 'pull' ions in against the concentration gradient

WS During your course, you may investigate the effect of salt or sugar solutions on plant tissue.

Here is one experiment you could do.
1. Immerse raw potato cut into chips of equal length in sugar solutions of various concentration.
2. You will see the potato chips change in length depending on whether individual cells have lost or gained water.

Can you **predict** what would happen to the potato chips immersed in:
➤ concentrated sugar (e.g. 1 molar)
➤ medium concentration sugar (e.g. 0.5 molar)
➤ water (0 molar)?

Keywords

Surface area to volume ratio ➤ A number calculated by dividing the total surface area of an object by its volume. When the ratio is **high**, the efficiency of diffusion and other processes is **greater**

Water potential/diffusion gradient ➤ A higher concentration of particle numbers in one area than another; in living systems, these areas are often separated by a membrane or cell wall

Partially permeable membrane ➤ A membrane with microscopic holes that allows small particles through (e.g. water) but not large ones (e.g. sugar)

1. Which has the highest surface area to volume ratio – an elephant or a shrew?
2. Some plant tissue is placed in a highly concentrated salt solution. Explain why water leaves the cells.

A plant's system is made up of organs and tissues that enable it to be a **photosynthetic organism**.

These are the main plant structures and their functions.

➤ **Roots** absorb water and minerals. They anchor plants in the soil.
➤ The **stem** transports water and nutrients to leaves. It holds leaves up to the light for maximum absorption of energy.
➤ The **leaf** is the organ of photosynthesis.
➤ The **flower** makes sexual reproduction possible through pollination.

Plant tissues, organs and systems

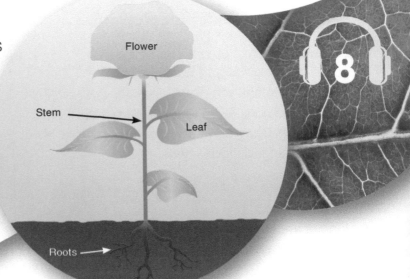

Leaves

As this cross-section of a leaf shows, leaf tissues are adapted for efficient photosynthesis. The epidermis covers the upper and lower surfaces of the leaf and protects the plant against pathogens.

Upper epidermis – cells are thin and flat to allow light to pass through

Palisade layer (mesophyll) – contains many chloroplasts for light absorption. It is positioned near the top of the leaf to be nearer to sunlight

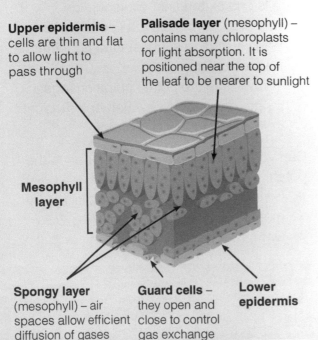

Mesophyll layer

Spongy layer (mesophyll) – air spaces allow efficient diffusion of gases

Guard cells – they open and close to control gas exchange

Lower epidermis

Stem and roots

Veins in the stem, roots and leaves contain tissues that transport water, carbohydrate and minerals around the plant.

➤ **Xylem tissue** transports water and mineral ions from the roots to the rest of the plant.
➤ **Phloem tissue** transports dissolved sugars from the leaves to the rest of the plant.

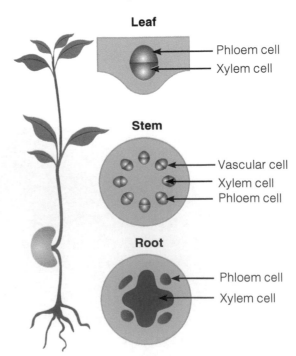

Leaf
— Phloem cell
— Xylem cell

Stem
— Vascular cell
— Xylem cell
— Phloem cell

Root
— Phloem cell
— Xylem cell

Meristem tissue is found at the growing tips of shoots and roots.

Xylem, phloem and root hair cells

Xylem, phloem and root hair cells are adapted to their function.

Part of plant	Appearance	Function	How they are adapted to their function
Xylem	Hollow tubes made from dead plant cells (the hollow centre is called a lumen)	Transport water and mineral ions from the roots to the rest of the plant in a process called **transpiration**	The cellulose cell walls are thickened and strengthened with a waterproof substance called **lignin**
Phloem	Columns of living cells	**Translocate** (move) cell sap containing sugars (particularly sucrose) from the leaves to the rest of the plant, where it is either used or stored	Phloem have pores in the end walls so that the cell sap can move from one phloem cell to the next
Root hair cells	Long and thin; have hair-like extensions	Absorb minerals and water from the soil	Large surface area

Keywords

Photosynthetic organism ➤ Able to absorb light energy and manufacture carbohydrate from carbon dioxide and water

Transpiration ➤ Flow of water through the plant ending in evaporation from leaves

Lignin ➤ Strengthening, waterproof material found in walls of xylem cells

Translocation ➤ Process in which sugars move through the phloem

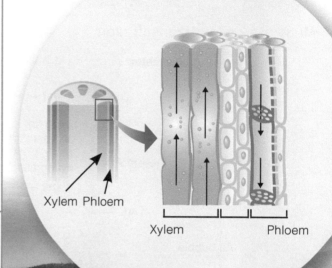

Xylem Phloem

Xylem   Phloem

If you snap a stick of celery open and pull the two halves apart, long 'strings' will pull away at the break. These are the long vascular tubes that carry water up to the celery's leaves.

➤ Cut a section from a stick of celery and look at it end-on. Can you see the vascular bundles that contain the xylem and phloem? To see the section more clearly, use a hand lens or magnifying glass.

➤ Try putting a celery stick (bottom end down) in a cup of food colouring overnight. When you cut the stem open the next day, you will see the coloured dye in the vascular bundles.

1. What part of the plant organ system allows water to enter the plant?
2. Why are there gaps between cells in the spongy mesophyll?
3. Why do xylem cells not have end walls?

Transport in plants

Transpiration

The movement of water through a plant, from roots to leaves, takes place as a transpiration stream. Once water is in the leaves, it diffuses out of the stomata into the surrounding air. This is called **(evapo)transpiration**.

Water evaporates from the spongy mesophyll through the stomata.	Water passes by osmosis from the xylem vessels in the leaf into the spongy mesophyll cells to replace what has been lost.	This movement 'pulls' the column of water in that xylem vessel upwards.	Water enters root hair cells by osmosis to replace water that has entered the xylem.

Measuring rate of transpiration

A potometer

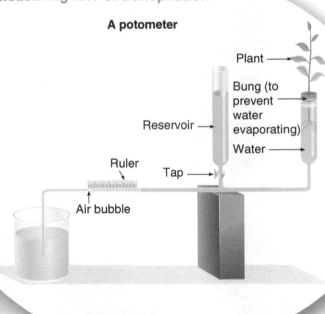

A leafy shoot's rate of transpiration can be measured using a **potometer**.

The shoot is held in a tube with a bung around the top to prevent any water from evaporating (this would give a false measurement of the water lost by transpiration).

As the plant transpires, it takes up water from the tube to replace what it has lost. All the water is then pulled up, moving the air bubble along.

The distance the air bubble moves can be used to calculate the plant's rate of transpiration for a given time period.

The experiment can be repeated, varying a different factor each time, to see how each factor affects the rate of transpiration.

Factors affecting rate of transpiration

Evaporation of water from the leaf is affected by **temperature**, **humidity**, **air movement** and **light intensity**.

➤ **Increased temperature** increases the kinetic energy of molecules and removes water vapour more quickly.

➤ **Increased air movement** removes water vapour molecules.

➤ **Increased light intensity** increases the rate of photosynthesis. This in turn draws up more water from the transpiration stream, which maintains high concentration of water in the spongy mesophyll.

➤ **Decreasing atmospheric humidity** lowers water vapour concentration outside of the stoma and so maintains the concentration gradient.

How water vapour exits the leaf

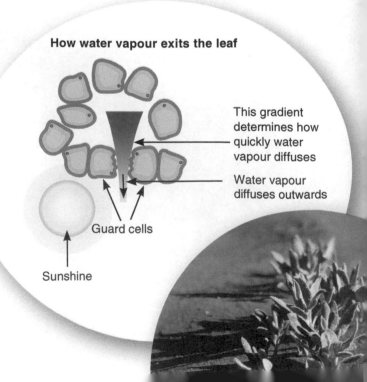

This gradient determines how quickly water vapour diffuses

Water vapour diffuses outwards

Guard cells

Sunshine

Opening and closing of stomata

Guard cells control the amount of water vapour that evaporates from the leaves and the amount of carbon dioxide that enters them.

➤ When light intensity is high and photosynthesis is taking place at a rapid rate, the sugar concentration rises in photosynthesising cells, e.g. palisade and guard cells.

➤ Guard cells respond to this by increasing the rate of water movement in the transpiration stream. This in turn provides more water for photosynthesis.

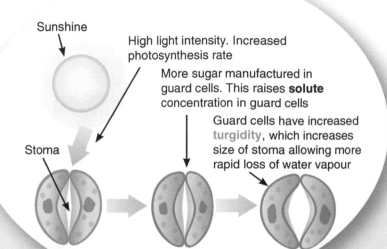

Sunshine

High light intensity. Increased photosynthesis rate

More sugar manufactured in guard cells. This raises **solute** concentration in guard cells

Guard cells have increased turgidity, which increases size of stoma allowing more rapid loss of water vapour

Stoma

Stomata

(WS) A **hypothesis** is an idea or explanation that you test through study and experiments. It should include a reason. For example: desert plants have fewer stomata than temperate plants **because** they need to minimise water loss.

➤ In an experiment investigating the factors that affect the rate of transpiration, a student plans to take measurements of weight loss or gain from a privet plant.

➤ Construct hypotheses for each of these factors: **temperature**, **humidity**, **air movement** and **light intensity**.

The first has been done for you: As temperature increases, the plant will lose mass/water more quickly **because** diffusion occurs more rapidly.

➤ Paint a thin layer of clear nail varnish onto the underside of a waxy leaf such as laurel – about 2 cm² should be sufficient.

➤ Allow the varnish to dry for at least 15 minutes. This will create a mould of the stomata as the liquid varnish fills the pores in the leaf.

➤ Peel the varnish strip off. Take care not to allow it to fold over or roll.

➤ Prepare it on a microscope slide with a drop of water and a cover slip.

➤ View it under medium power and you should see the stomata.

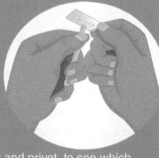

Compare different types of leaf, e.g. holly and privet, to see which leaves have most stomata. If you don't have a microscope at home, ask to use one at school.

1. What effect would **decreasing** air humidity have on transpiration?
2. In what circumstances might it be beneficial for plants to **close** their stomata?
3. How does water pass from xylem vessels into the leaf?
4. Describe how water passes from roots to leaves.

Keywords

(Evapo)transpiration ➤ Evaporation of water from stomata in the leaf

Turgidity ➤ Where plant cells fill with water and swell as a result of osmosis

Transport in humans 1

🎧 **10**

Blood circulation

Blood moves around the body in a **double circulatory system**. In other words, blood moves twice through the heart for every full circuit. This ensures maximum efficiency for absorbing oxygen and delivering materials to all living cells.

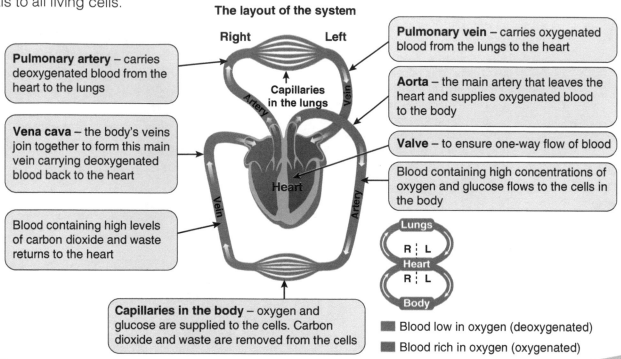

The layout of the system

Right   Left

Pulmonary artery – carries deoxygenated blood from the heart to the lungs

Capillaries in the lungs

Artery   Vein

Vena cava – the body's veins join together to form this main vein carrying deoxygenated blood back to the heart

Heart

Vein   Artery

Blood containing high levels of carbon dioxide and waste returns to the heart

Capillaries in the body – oxygen and glucose are supplied to the cells. Carbon dioxide and waste are removed from the cells

Pulmonary vein – carries oxygenated blood from the lungs to the heart

Aorta – the main artery that leaves the heart and supplies oxygenated blood to the body

Valve – to ensure one-way flow of blood

Blood containing high concentrations of oxygen and glucose flows to the cells in the body

Lungs
R ⁞ L
Heart
R ⁞ L
Body

■ Blood low in oxygen (deoxygenated)
■ Blood rich in oxygen (oxygenated)

The heart

The heart is made of powerful muscles that contract and relax rhythmically in order to continuously pump blood around the body. The **heart muscle** is supplied with food (particularly glucose) and oxygen through the **coronary artery**.

The sequence of events that takes place when the heart beats is called the **cardiac cycle**.

1. The heart relaxes and blood enters both atria from the veins.
2. The atria contract together to push blood into the ventricles, opening the atrioventricular valves.
3. The ventricles contract from the bottom, pushing blood upwards into the arteries. The backflow of blood into the ventricles is prevented by the **semilunar valves**.

The left side of the heart is more muscular than the right because it has to pump blood further round the body. The right side only has to pump blood to the lungs and back.

A useful measurement for scientists and doctors to take is **cardiac output**. This is calculated using:

cardiac output = stroke volume × heart rate

So, for a person who pumps out 70 ml of blood in one heartbeat (stroke volume) and has a pulse of 70 beats per minute, the cardiac output would be 4900 ml per minute.

Controlling the heartbeat

The heart is stimulated to beat rhythmically by pacemaker cells. The pacemaker cells produce impulses that spread across the atria to make them contract. Impulses are spread from here down to the ventricles, making them contract, pushing blood up and out.

Nerves connecting the heart to the brain can increase or decrease the pace of the pacemaker cells in order to regulate the heartbeat.

If a person has an irregular heartbeat, they can be fitted with an artificial, electrical pacemaker.

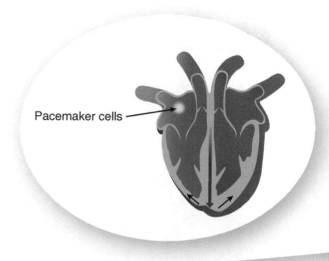

Pacemaker cells

Blood vessels

Blood is carried through the body in three types of vessel.

➤ **Arteries** have thick walls made of elastic fibres and muscle fibres to cope with the high pressure. The **lumen** (space inside) is small compared to the thickness of the walls. There are no valves.

➤ **Veins** have thinner walls. The lumen is much bigger compared to the thickness of the walls and there are valves to prevent the backflow of blood.

➤ **Capillaries** are narrow vessels with walls only one cell thick. These microscopic vessels connect arteries to veins, forming dense networks or **beds**. They are the only blood vessels that have permeable walls to allow the exchange of materials.

Artery

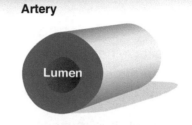

Vein

Valve

Capillary

Note: capillaries are much smaller than veins or arteries

Coronary heart disease

Coronary heart disease (CHD) is a **non-communicable** disease. It results from the build-up of **cholesterol**, leading to **plaques** laid down in the coronary arteries. This restricts blood flow and the artery may become blocked with a blood clot or **thrombosis**. The heart muscle is deprived of glucose and oxygen, which causes a **heart attack**.

The likelihood of plaque developing increases if you have a high fat diet. The risk of having a heart attack can be reduced by:

➤ eating a balanced diet and not being overweight
➤ not smoking tobacco
➤ lowering alcohol intake
➤ reducing salt levels in your diet
➤ reducing stress levels.

Healthy artery

Build-up of fatty material begins

Plaque forms

Plaque ruptures; blood clot forms

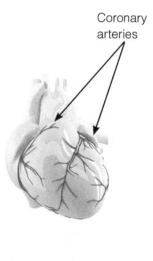

Coronary arteries

1. What is the function of valves in veins?
2. Describe how eating a diet high in fat can lead to a heart attack.

Transport in humans 2

Keywords

Haemoglobin ➤ Iron-containing molecule that binds to oxygen molecules in red blood cells

Ventilation ➤ Process of drawing air into and out of the lungs. It involves the ribs, intercostal muscles and diaphragm

Remedying heart disease

For patients who have heart disease, artificial implants called **stents** can be used to increase blood flow through the coronary artery.

Statins are a type of drug that can be taken to reduce blood cholesterol levels.

In some people, the heart valves may deteriorate, preventing them from opening properly. Alternatively, the valve may develop a leak.

This means that the supply of oxygenated blood to vital organs is reduced. The problem can be corrected by surgical replacement using a **biological** or **mechanical valve**.

When complete heart failure occurs, a heart transplant can be carried out. If a donor heart is unavailable, the patient may be kept alive by an artificial heart until one can be found. Mechanical hearts are also used to give the biological heart a rest while it recovers.

Blood as a tissue

Blood transports digested food and oxygen to cells and removes the cells' waste products. It also forms part of the body's defence mechanism.

The four components of blood are:
➤ platelets
➤ plasma
➤ white blood cells
➤ red blood cells.

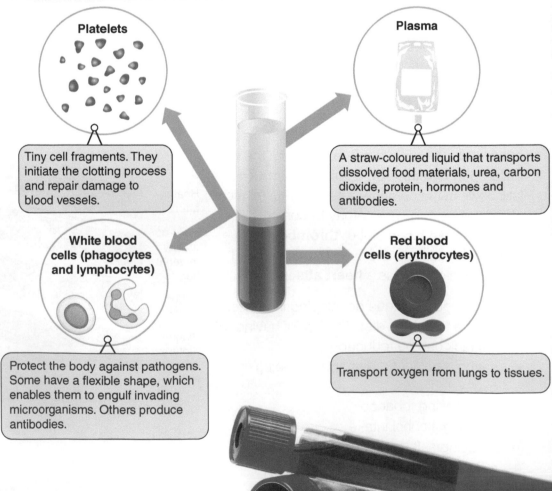

Platelets

Tiny cell fragments. They initiate the clotting process and repair damage to blood vessels.

Plasma

A straw-coloured liquid that transports dissolved food materials, urea, carbon dioxide, protein, hormones and antibodies.

White blood cells (phagocytes and lymphocytes)

Protect the body against pathogens. Some have a flexible shape, which enables them to engulf invading microorganisms. Others produce antibodies.

Red blood cells (erythrocytes)

Transport oxygen from lungs to tissues.

Oxygen transport

Red blood cells are small and have a biconcave shape. This gives them a large surface area to volume ratio for absorbing oxygen. When the cells reach the lungs, they absorb and bind to the oxygen in a molecule called **haemoglobin**.

> haemoglobin + oxygen $\rightleftharpoons$ oxyhaemoglobin

Blood is then pumped around the body to the tissues, where the reverse of the reaction takes place. Oxygen diffuses out of the red blood cells and into the tissues.

Transport of oxygen in red blood cells

Oxygen from alveolus — Red blood cell

Haemoglobin molecules

Oxygen released to cells

The lungs

Humans, like many vertebrates, have lungs to act as a **gaseous exchange surface**.

Other structures in the **thorax** enable air to enter and leave the lungs (**ventilation**).

➤ The **trachea** is a flexible tube, surrounded by rings of cartilage to stop it collapsing. Air is breathed in via the mouth and passes through here on its way to the lungs.

➤ **Bronchi** are branches of the trachea.

➤ The **alveoli** are small air sacs that provide a large surface area for the exchange of gases.

➤ **Capillaries** form a dense network to absorb maximum oxygen and release carbon dioxide.

In the alveoli, **oxygen** diffuses down a concentration gradient. It moves across the thin layers of cells in the alveolar and capillary walls, and into the red blood cells.

For **carbon dioxide**, the gradient operates in reverse. The carbon dioxide passes from the blood to the alveoli, and from there it travels back up the air passages to the mouth.

The lungs

Trachea (windpipe)
Lung
Bronchiole
Pleural membrane
Bronchus (bronchi)
Alveolus (alveoli)

A single alveolus and a capillary

Deoxygenated blood
CO_2
Oxygenated blood
CO_2
O_2
Alveolus
O_2

WS Scientists make observations, take measurements and gather data using a variety of instruments and techniques. Recording data is an important skill.

Create a table template that you could use to record data for the following experiment:

An investigation that involves measuring the resting and active pulse rates of 30 boys and 30 girls, together with their average breathing rates.

Make sure that:

➤ you have the correct number of columns and rows

➤ each variable is in a heading

➤ units are in the headings (so they don't need to be repeated in the body of the table).

1. What are the differences between lymphocytes and red blood cells?
2. Describe the route that would be travelled by a molecule of oxygen through the body until it reached a respiring muscle cell. State the cells, tissue, organs and processes that are involved.

Plants are **producers**. This means they can photosynthesise, i.e. make food molecules in the form of **carbohydrate** from the simple molecules, carbon dioxide and water. As such, they are the main producers of **biomass**. Sunlight energy is needed for photosynthesis.

Photosynthesis

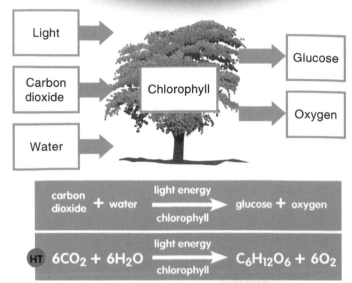

Photosynthesis:
➤ is an **endothermic** reaction
➤ requires **chlorophyll** to absorb the sunlight; this is found in the **chloroplasts** of photosynthesising cells, e.g. palisade cells, guard cells and spongy mesophyll cells
➤ produces **glucose**, which is then respired for energy release or converted to other useful molecules for the plant
➤ produces **oxygen** that has built up in the atmosphere over millions of years; oxygen is vital for respiration in all organisms.

$$\text{carbon dioxide} + \text{water} \xrightarrow[\text{chlorophyll}]{\text{light energy}} \text{glucose} + \text{oxygen}$$

HT $$6CO_2 + 6H_2O \xrightarrow[\text{chlorophyll}]{\text{light energy}} C_6H_{12}O_6 + 6O_2$$

Rate of photosynthesis

The rate of photosynthesis can be affected by:
➤ temperature
➤ light intensity
➤ carbon dioxide concentration
➤ amount of chlorophyll.

In a given set of circumstances, **temperature**, **light intensity** and **carbon dioxide concentration** can act as **limiting factors**.

Temperature

1 As the temperature rises, so does the rate of photosynthesis. This means temperature is limiting the rate of photosynthesis.

2 As the temperature approaches 45°C, the enzymes controlling photosynthesis start to be denatured. The rate of photosynthesis decreases and eventually declines to zero.

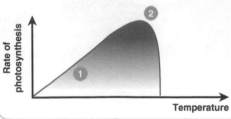

Light intensity

1 As the light intensity increases, so does the rate of photosynthesis. This means light intensity is limiting the rate of photosynthesis.

2 Eventually, the rise in light intensity has no effect on photosynthesis rate. Light intensity is no longer the limiting factor; carbon dioxide or temperature must be.

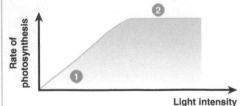

Carbon dioxide concentration

1 As carbon dioxide concentration increases, so does the rate of photosynthesis. Carbon dioxide concentration is the limiting factor.

2 Eventually, the rise in carbon dioxide concentration has no effect – it is no longer the limiting factor.

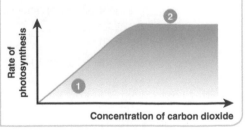

You need to understand that each factor has the potential to increase the rate of photosynthesis.

HT You also need to explain how these factors interact in terms of which variable is acting as the limiting factor.

The inverse law

The effect of light intensity on photosynthesis can be investigated by placing a lamp at varying distances from a plant. As the lamp is moved further away from the plant the light intensity decreases, as shown in the graph.

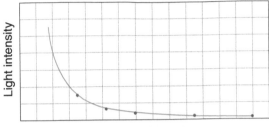

Distance from plant

There is an **inverse relationship** between the two variables. The graph can be used to convert distances to light intensity, or light intensity can be calculated using the formula:

$$\text{light intensity} = \frac{1}{d^2}$$

> d is the distance from the lamp.

There are no units of light intensity – it has an arbitrary scale.

Commercial applications

Farmers and market gardeners can increase their crop yields in greenhouses. They do this by:

> making the temperature optimum for growth using heaters
> increasing light intensity using lamps
> installing fossil-fuel burning stoves to increase carbon dioxide concentration (and increase temperature).

If applied carefully, the cost of adding these features will be offset by increased profit from the resulting crop.

Keywords

Biomass ➤ Mass of organisms calculated by multiplying their individual mass by the number that exist

Endothermic ➤ A change that requires the input of energy

Chlorophyll ➤ A molecule that gives plants their green colour and absorbs light energy

Limiting factor ➤ A variable that, if changed, will influence the rate of reaction most

Cellulose ➤ Large carbohydrate molecule found in all plants; an essential constituent of cell walls

Uses of glucose in plants

The glucose produced from photosynthesis can be used immediately in respiration, but some is used to synthesise larger molecules: **starch**, **cellulose**, **protein** and **lipids**.

Starch is insoluble. So it is suitable for storage in leaves, stems or roots.

Cellulose is needed for cell walls.

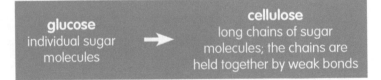

Protein is used for the growth and repair of plant tissue, and also to synthesise enzyme molecules.

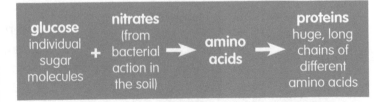

Lipids are needed in cell membranes, and for fat and oil storage in seeds.

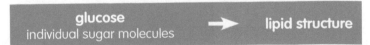

Use plasticine and coloured paper / card to produce a molecular model showing the photosynthesis reaction. Use ideas from the 'Uses of glucose in plants' section to help you.

🎧 **12**

✓

1. Give two reasons why photosynthesis is seen as the opposite of respiration.
HT 2. What **economic** factor must market gardeners consider before installing wood-burning stoves in their greenhouses?

Mind map

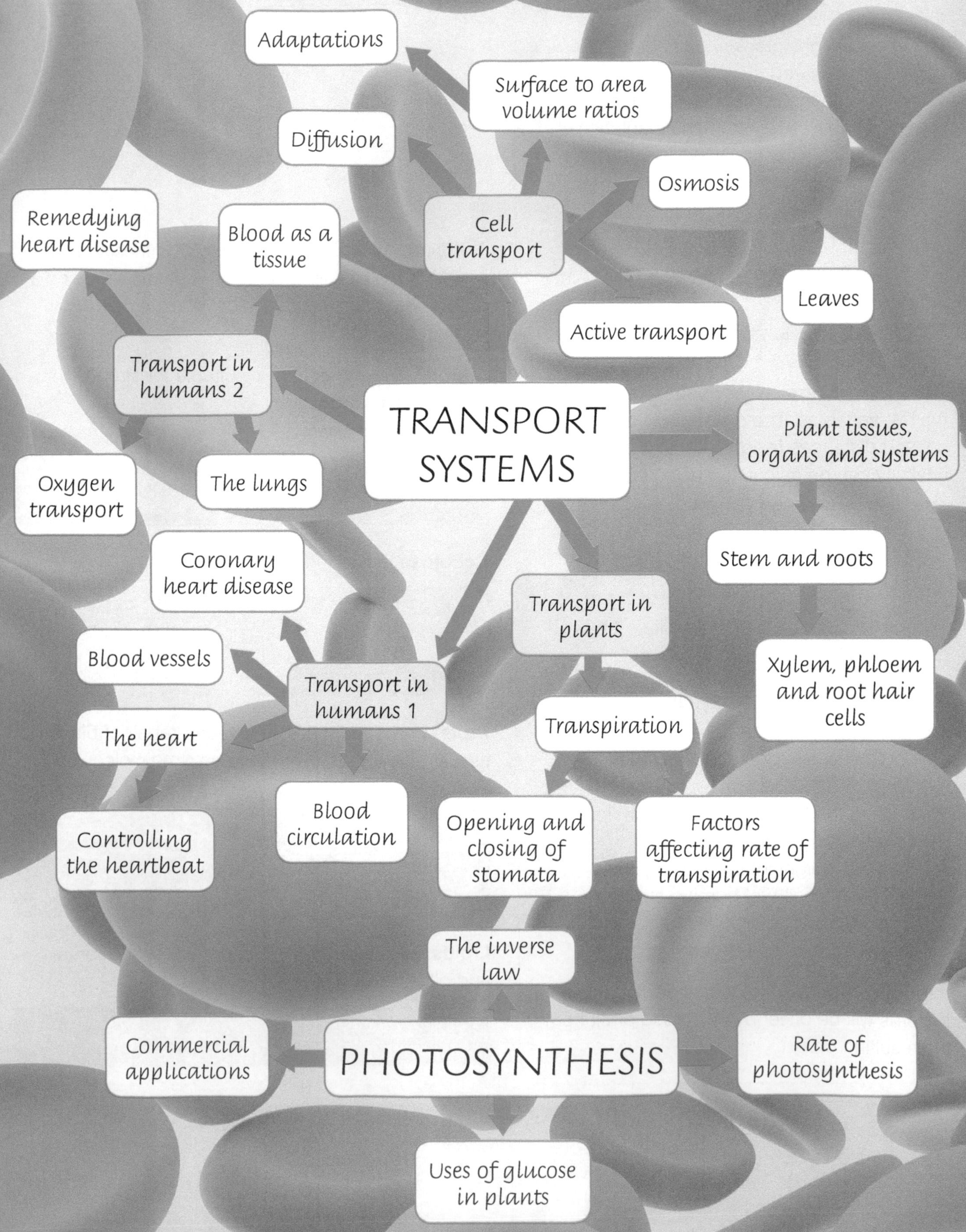

Adaptations

Surface to area
volume ratios

Diffusion

Osmosis

Remedying
heart disease

Blood as a
tissue

Cell
transport

Leaves

Active transport

Transport in
humans 2

TRANSPORT
SYSTEMS

Plant tissues,
organs and systems

Oxygen
transport

The lungs

Stem and roots

Coronary
heart disease

Transport in
plants

Blood vessels

Transport in
humans 1

Xylem, phloem
and root hair
cells

The heart

Transpiration

Controlling
the heartbeat

Blood
circulation

Opening and
closing of
stomata

Factors
affecting rate of
transpiration

The inverse
law

Commercial
applications

PHOTOSYNTHESIS

Rate of
photosynthesis

Uses of glucose
in plants

Practice questions

1. Which of the following are examples of osmosis? Tick (✓) the three correct options. **(3 marks)**

   a) Water evaporating from leaves ☐

   b) Water moving from plant cell to plant cell and back again ☐

   c) Mixing pure water and sugar solution ☐

   d) A pear submerged in a concentrated sugar solution losing water ☐

   e) Water moving from blood plasma to body cells ☐

   f) Sugar being absorbed from the intestine into the blood ☐

2. Emphysema is a lung disease that increases the thickness of the surface of the lungs for gas exchange and reduces the total area available for gas exchange.

   Two men did the same amount of exercise. One man was in good health and the other man had emphysema.

   The results are shown in the table.

	Healthy man	Man with emphysema
Total air flowing into lungs (dm³/min)	89.5	38.9
Oxygen entering blood (dm³/min)	2.5	1.2

   a) Which man had more oxygen entering his blood? **(1 mark)**

   b) Explain why the man with emphysema struggled to carry out exercise. **(2 marks)**

3. The diagram shows two types of blood vessel.

   A   **B**

    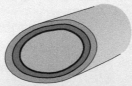

   a) Name each type of blood vessel. **(2 marks)**

   b) Explain why blood vessel A has a thick, elastic muscle wall. **(1 mark)**

   c) Why does blood vessel B have valves? **(1 mark)**

Non-communicable diseases

Keywords

Pathogen ➤ A microorganism that causes disease

Symptoms ➤ Physical or mental features that indicate a condition or disease, e.g. rash, high temperature, vomiting

Immune system ➤ A system of cells and other components that protect the body from pathogens and other foreign substances

Malnutrition ➤ A diet lacking in one or more food groups

Communicable diseases are caused by **pathogens** such as bacteria and viruses. They can be transmitted from organism to organism in a variety of ways. Examples include cholera and tuberculosis.

Non-communicable diseases are not primarily caused by pathogens. Examples are diseases caused by a poor diet, diabetes, heart disease and smoking-related diseases.

Health is the state of physical, social and mental well-being. Many factors can have an effect on health, including stress and life situations.

Risk factors

Non-communicable diseases often result from a combination of several **risk factors**.

➤ Risk factors produce an increased likelihood of developing that particular disease. They can be aspects of a person's lifestyle or substances found in the body or environment.

➤ Some of these factors are difficult to quantify or to establish as a definite **causal connection**. So scientists have to describe their effects in terms of probability or likelihood.

The **symptoms** observed in the body may result from communicable and non-communicable components interacting.

A lowered **immune system** may make a person more vulnerable to infection.

Immune reactions caused by pathogens can trigger allergies such as asthma and skin rashes.

Symptoms

Viruses inhabiting living cells can change them into cancer cells.

Serious physical health problems can lead to **mental illness** such as depression.

Poor diet

People need a **balanced diet**.

If a diet does not include enough of the main food groups, **malnutrition** might result. Lack of correct vitamins leads to diseases such as **scurvy** and **rickets**. Lack of the mineral, iron, results in **anaemia**.

A high fat diet contributes to cardiovascular disease and high levels of salt increase blood pressure.

Smoking tobacco

Chemicals in tobacco smoke affect health.

➤ **Carbon monoxide** decreases the blood's oxygen-carrying capacity.

➤ **Nicotine** raises the heart rate and therefore blood pressure.

➤ **Tar** triggers cancer.

➤ **Particulates** cause **emphysema** and increase the likelihood of **lung infections**.

Weight/lack of exercise

Obesity and lack of exercise both increase the risk of developing **type 2 diabetes** and cardiovascular disease.

One way to show whether someone is underweight or overweight for their height is to calculate their **body mass index (BMI)**, using the following formula:

$$BMI = \frac{mass\ (kg)}{height^2\ (m)}$$

Recommended BMI chart

BMI	What it means
<18.5	Underweight – too light for your height
18.5–25	Ideal – correct weight range for your height
25–30	Overweight – too heavy
30–40	Obese – much too heavy. Health risks!

Example:

Calculate a man's BMI if he is 1.65 m tall and weighs 68 kg.

$$BMI = \frac{mass\ (kg)}{height^2\ (m)} = \frac{68}{1.65^2} = \frac{68}{2.7} = 25$$

The recommended BMI for his height (1.65 m) is 18.5–25, so he is just a healthy weight.

There are drawbacks to using BMI as a way of assessing people's health. For example:

➤ teenagers go through a rapid growth phase
➤ a person could have a well-developed muscle system – this would increase their body mass but not make them obese.

Some scientists say a more accurate method is using the waist/hip ratio. A tape measure is used to measure the circumference of the hips and the waist (at its widest). The following chart can then be used.

Waist to hip ratio (WHR)		
Male	Female	Health risk based solely on WHR
0.95 or below	0.80 or below	Low
0.96 to 1.0	0.81 to 0.85	Moderate
1.0+	0.85+	High

Alcohol

Drinking excess alcohol can impair brain function and lead to **cirrhosis** of the liver. It also contributes to some types of cancer and cardiovascular disease.

Smoking and drinking alcohol during pregnancy

Unborn babies receive nutrition from the mother via the placenta. Substances from tobacco, alcohol and other drugs can pass to the baby and cause **lower birth weight**, **foetal alcohol syndrome** and **addiction**.

Carcinogens

Exposure to **ionising radiation** (for example, X-rays, gamma rays) can cause cancerous tumours. Overexposure to UV light can cause skin cancer. Certain chemicals such as mercury can also increase the likelihood of cancer.

 Interpreting complex data in graphs doesn't need to be difficult. This line graph shows data about smoking and lung cancer. Look for different patterns in it. For example:

➤ males have higher smoking rates in all years
➤ female cancer rates have increased overall since 1972.

Can you see any other patterns?

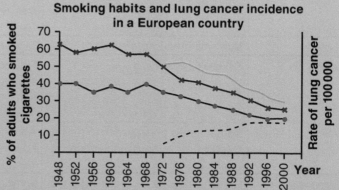

Smoking habits and lung cancer incidence in a European country

KEY
×–×–× Male smoking data — Male incidence of lung cancer
●–●–● Female smoking data – – – – Female incidence of lung cancer

1. State three factors that cause cancer.
2. Give one consequence of a lowered immune system.

Communicable diseases

How do pathogens spread?

Pathogens are disease-causing microorganisms from groups of bacteria, viruses, fungi and protists. All animals and plants can be affected by pathogens. They spread in many ways, including:

➤ **droplet infection** (sneezing and coughing), e.g. flu

➤ **physical contact**, such as touching a contaminated object or person, e.g. Ebola virus

➤ **transmission** by transferral of or contact **with bodily fluids**, e.g. hepatitis B

➤ **sexual transmission**, e.g. HIV, gonorrhoea

➤ **contamination of food or water**, e.g. Salmonella, cholera

➤ **animal bites**, e.g. rabies.

How do pathogens cause harm?

➤ Bacteria and viruses reproduce rapidly in the body.

➤ Viruses cause cell damage.

➤ Bacteria produce toxins that damage tissues.

These effects produce **symptoms** in the body.

How can the spread of disease be prevented?

The spread of disease can be prevented by:

➤ good hygiene, e.g. washing hands/whole body, using soaps and disinfectants

➤ destroying **vectors**, e.g. disrupting the life cycle of mosquitoes can combat malaria

➤ the isolation or quarantine of individuals

➤ vaccination.

Bacterial diseases

Disease	Transmission	Symptoms	Treatment/prevention
Tuberculosis	Droplet infection	Persistent coughing, which may bring up blood; chest pain; weight loss; fatigue; fever; night sweats; chills	Long course of antibiotics
Cholera	Contaminated water/food	Diarrhoea; vomiting; dehydration	Rehydration salts
Chlamydia	Sexually transmitted	May not be present, but can include discharge and bleeding from sex organs	Antibiotics Using condoms during sexual intercourse can reduce chances of infection
Helicobacter	Spread orally from person to person by saliva, or by fecal contamination	Abdominal pain; feeling bloated; nausea; vomiting; loss of appetite; weight loss	Proton pump inhibitors and antibiotics
Salmonella	Contaminated food containing toxins from pathogens – these could be introduced from unhygienic food preparation techniques	Vomiting; fever; diarrhoea, stomach cramps	Anti-diarrhoeals and antibiotics; vaccinations for chickens
Gonorrhoea	Sexually transmitted	Thick yellow or green discharge from vagina or penis; pain on urination	Antibiotic injection followed by antibiotic tablets; penicillin is no longer effective against gonorrhoea; prevention through use of condoms

Fungal diseases

Disease	Transmission	Symptoms	Treatment/prevention
Athlete's foot	Direct and indirect contact, e.g. skin-to-skin, bed sheets and towels (often spreads at swimming pools and in changing rooms)	Itchy, red, scaly, flaky and dry skin	Self-care and anti-fungal medication externally applied

Viral and protist diseases

Disease	Transmission	Symptoms and notes	Treatment/prevention
Measles (viral)	Droplets from sneezes and coughs	Fever; red skin rash; fatal if complications arise	No specific treatment; vaccine is a highly effective preventative measure
HIV (viral)	Sexually transmitted; exchange of body fluids; sharing of needles during drug use	Flu-like symptoms initially; late-stage AIDS produces complications due to compromised immune system	Anti-retroviral drugs
Ebola (viral)	Via body fluids; contaminated needles; bite from infected animal	Early stages: muscle pain; sore throat; diarrhoea Later stages: kidney/liver failure; internal bleeding	No direct treatments or tested vaccines are available at the time of writing; symptoms and infections are treated as they appear
Malaria (protist)	Via mosquito vector	Headache; sweats; chills and vomiting; symptoms disappear and reappear on a cyclical basis; further life-threatening complications may arise	Various anti-malarial drugs are available for both prevention and cure; prevention of mosquito breeding and use of mosquito nets

Viruses carry out their life cycle largely within a host cell. Viruses tend to follow one of two types of cycle.

➤ **Lysogenic**: the virus does not lyse (split) the host cell. Instead, the viral DNA becomes part of the host DNA. When the host cell divides, the viral DNA is also copied and is present in each of the daughter cells.

➤ **Lytic**: the virus replicates using the host DNA (as before). The new viruses that are formed destroy the infected cell and its membrane as they are released.

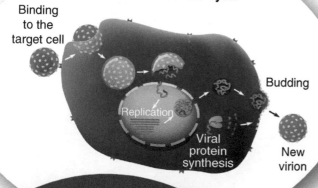

Influenza viral life cycle

Binding to the target cell

Replication

Viral protein synthesis

Budding

New virion

➤ Create cards showing information about the disease-causing microorganisms in this module. Each card should have the name of the disease at the top and then list the symptoms, mode of transmission, etc.

➤ Use the cards with a revision buddy. You could give a score for each category of information, e.g. contagion factor (how easily the disease is transmitted), severity of symptoms, etc.

1. Why is cholera transmitted rapidly in areas that have poor sanitation (sewage systems)?
2. Name two diseases that cause a rash or affect the skin in some way.

Keywords

Droplet infection ➤ Transmission of microorganisms through the aerosol (water droplets) produced through coughing and sneezing

Vectors ➤ Small organisms (such as mosquitoes or ticks) that pass on pathogens between people or places

Human defences

Keywords

Epithelial ➤ A single layer of cells often found lining respiratory and digestive structures

Mucus ➤ Thick fluid produced in the lining of respiratory and digestive passages

Cilia ➤ Microscopic hairs found on the surface of epithelial cells; they 'waft' from side to side in a rhythmic manner

Antigen ➤ Molecular marker on a pathogen cell membrane that acts as a recognition point for antibodies

Antibodies ➤ Proteins produced by white blood cells (particularly lymphocytes). They lock on to antigens and neutralise them

Immunological memory ➤ The system of cells and cell products whose function is to prevent and destroy microbial infection

Non-specific defences

The body has a number of general or non-specific defences to stop pathogens multiplying inside it.

The skin covers most of the body – it is a **physical barrier** to pathogens. It also secretes antimicrobial peptides to kill microorganisms. If the skin is damaged, a clotting mechanism takes place in the blood preventing pathogens from entering the site of the wound.

Tears contain enzymes called **lysozymes**. Lysozymes break down pathogen cells that might otherwise gain entry to the body through tear ducts.

Hairs in the **nose** trap particles that may contain pathogens.

Tubes in the respiratory system (**trachea** and **bronchi**) are lined with special **epithelial** cells. These cells either produce a sticky, liquid **mucus** that traps microorganisms or have tiny hairs called **cilia** that move the mucus up to the mouth where it is swallowed.

The **stomach** produces **hydrochloric acid**, which kills microorganisms.

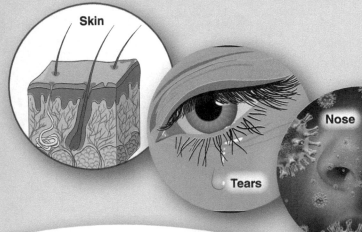

Skin

Tears

Nose

Epithelial cells

Cilia

Mucus produced here

Stomach

Phagocytes are a type of **white blood cell**. They move around in the bloodstream and body tissues searching for pathogens. When they find pathogens, they **engulf** and digest them in a process called **phagocytosis**.

White blood cell (phagocyte)

Microorganisms invade the body

The white blood cell surrounds and ingests the microorganisms

The white blood cell starts to digest the microorganisms

The microorganisms have been digested by the white blood cell

Specific defences

White blood cells called **lymphocytes** recognise molecular markers on pathogens called **antigens**. They produce **antibodies** that lock on to the antigens on the cell surface of the pathogen cell. The immobilised cells are clumped together and engulfed by phagocytes.

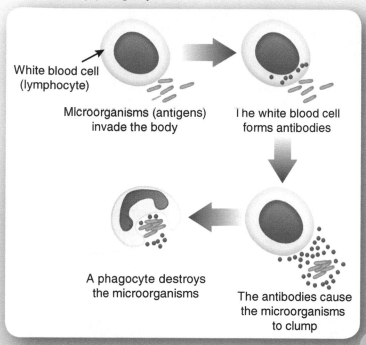

White blood cell (lymphocyte)

Microorganisms (antigens) invade the body

The white blood cell forms antibodies

A phagocyte destroys the microorganisms

The antibodies cause the microorganisms to clump

Some white blood cells produce **antitoxins** that neutralise the poisons produced from some pathogens.

Every pathogen has its own unique antigens. Lymphocytes make antibodies specifically for a particular antigen.

Example: Antibodies to fight TB will not fight cholera

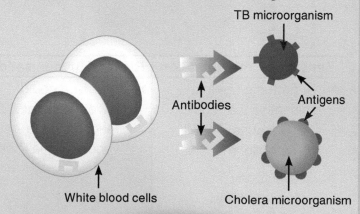

TB microorganism

Antibodies

Antigens

White blood cells

Cholera microorganism

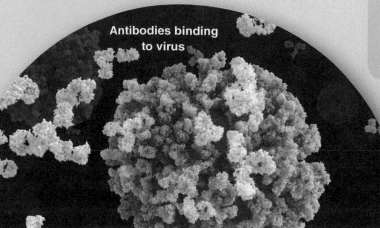

Antibodies binding to virus

Active immunity

Once lymphocytes recognise a particular pathogen, the interaction is stored as part of the body's **immunological memory** through **memory lymphocytes**. These memory cells can produce the right antibodies much quicker if the same pathogen is detected again, therefore providing future protection against the disease. The process is called the **secondary response** and is part of the body's **active immunity**. Active immunity can also be achieved through vaccination.

Memory lymphocytes and antibody production

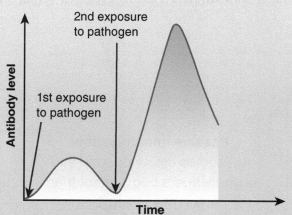

2nd exposure to pathogen

1st exposure to pathogen

Antibody level

Time

(WS) Investigating the growth of pathogens in the laboratory involves culturing microorganisms (see Module 3). This presents hazards that require a **risk assessment**. A risk assessment involves taking into account the severity of each hazard and the likelihood that it will occur.

Any experiment of this type involves thinking about risks in advance. Here is an example of a risk assessment table for this investigation.

Hazard	Infection from pathogen	Scald from autoclave (a specialised pressure cooker for superheating its contents)
Risk	High	High
How to lower the risk	➤ Observe **aseptic technique**. ➤ Wash hands thoroughly before and after experiment. ➤ Store plates at a maximum temperature of 25°C.	➤ Ensure lid is tightly secured. ➤ Adjust heat to prevent too high a pressure. ➤ Wait for autoclave to cool down before removing lid.

1. Describe the process of phagocytosis.
2. Explain how active immunity can be developed in the body.

Fighting disease

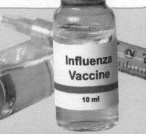

Keywords

Sensitisation ➤ Cells in the immune system are able to 'recognise' antigens or foreign cells and respond by attacking them or producing antibodies

Herd immunity ➤ Vaccination of a significant portion of a population (or herd) makes it hard for a disease to spread because there are so few people left to infect. This gives protection to susceptible individuals such as the elderly or infants

Inhibition ➤ The effect of one agent against another in order to slow down or stop activity, e.g. chemical reactions can be slowed down using inhibitors. Some hormones are inhibitors

Vaccination

There are two types of vaccination.

Passive immunisation

Antibodies are introduced into an individual's body, rather than the person producing them on their own. Some pathogens or toxins (e.g. snake venom) act very quickly and a person's immune system cannot produce antibodies quickly enough. So the person must be injected with the antibodies. However, this does not give long-term protection.

Active immunisation

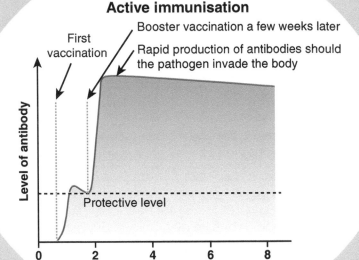

- First vaccination
- Booster vaccination a few weeks later
- Rapid production of antibodies should the pathogen invade the body
- Protective level

(Graph: Level of antibody vs Time (months), x-axis 0, 2, 4, 6, 8)

Immunisation gives a person immunity to a disease without the pathogens multiplying in the body, or the person having symptoms.

1. A weakened or inactive strain of the pathogen is injected. The pathogen is heat-treated so it cannot multiply. The antigen molecules remain intact.
2. Even though they are harmless, the antigens on the pathogen trigger the white blood cells to produce specific antibodies.
3. As with natural immunity, **memory lymphocytes** remain **sensitised**. This means they can produce more antibodies very quickly if the same pathogen is detected again.

Benefits of immunisation	Risks of immunisation
➤ It protects against diseases that could kill or cause disability (e.g. polio, measles).	➤ A person could have an allergic reaction to the vaccine (small risk).
➤ If everybody is vaccinated and **herd immunity** is established, the disease eventually dies out (this is what happened to smallpox).	

Influenza Vaccine 10 ml

16

HT

Antibiotics and painkillers

Diseases caused by bacteria (not viruses) can be treated using **antibiotics**, e.g. penicillin. Antibiotics are drugs that destroy the pathogen. Some bacteria need to be treated with antibiotics specific to them.

Antibiotics work because they **inhibit** cell processes in the bacteria but not the body of the host.

Viral diseases can be treated with **antiviral drugs**, e.g. swine flu can be treated with 'Tamiflu' tablets. It is a challenge to develop drugs that destroy viruses without harming body tissues.

Antibiotic resistance

Antibiotics are very effective at killing bacteria. However, some bacteria are **naturally resistant** to particular antibiotics. It is important for patients to follow instructions carefully and take the full course of antibiotics so that all the harmful bacteria are killed.

If doctors over-prescribe antibiotics, there is more chance of resistant bacteria surviving. These multiply and spread, making the antibiotic useless. **MRSA** is a bacterium that has become resistant to most antibiotics. These bacteria have been called 'superbugs'.

Painkillers

Painkillers or **analgesics** are given to patients to relieve symptoms of a disease, but they do not kill pathogens. Types of painkiller include paracetamol and ibuprofen. Morphine is another painkiller – it is a medicinal form of heroin used to treat extreme pain.

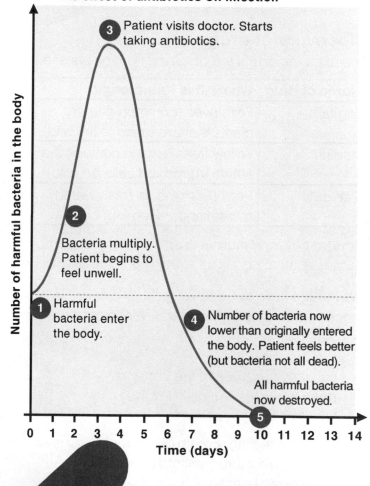

The effect of antibiotics on infection

Number of harmful bacteria in the body (y-axis)
Time (days) (x-axis)

3 — Patient visits doctor. Starts taking antibiotics.

2 — Bacteria multiply. Patient begins to feel unwell.

1 — Harmful bacteria enter the body.

4 — Number of bacteria now lower than originally entered the body. Patient feels better (but bacteria not all dead).

All harmful bacteria now destroyed.

5

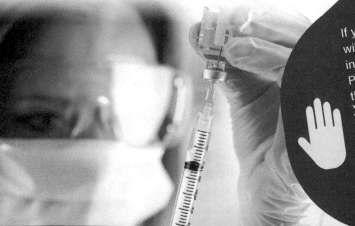

If you are revising with friends, try doing a role-play. This will help you visualise the actions of cells and processes in the immune system.

Plan and rehearse a scenario where a pathogen enters the human body and multiplies.

▶ Allocate roles for the pathogens and the various immune components: phagocytes, lymphocytes, antibodies, etc. You could even have a narrator.

▶ Mimic the actions of the various cells. How could you do this effectively?

▶ Ask someone to watch the role-play and give you feedback on what went well and what you could improve.

1. What is the difference between an antibiotic and an antibody?
2. What is the difference between an antiviral and an analgesic?
3. What can doctors and patients do to reduce the risk of antibiotic-resistant bacteria developing?

Drugs used for treating illnesses and health conditions include antibiotics, analgesics and other chemicals that modify body processes and chemical reactions. In the past, these drugs were obtained from plants and microorganisms.

Discovery and development of drugs

17

Discovery of drugs

The following drugs are obtained from plants and microorganisms.

Name of drug	Where it is found/origin	Use
Digitalis	Foxgloves (common garden plants that are found in the wild)	Slows down the heartbeat; can be used to treat heart conditions
Aspirin	Willow trees (aspirin contains the **active ingredient** salicylic acid)	Mild painkiller
Penicillin	Penicillium mould (discovered by Alexander Fleming)	Antibiotic

Modern **pharmaceutical drugs** are synthesised by chemists in laboratories, usually at great cost. The starting point might still be a chemical extracted from a plant.

New drugs have to be developed all the time to combat new and different diseases. This is a lengthy process, taking up to ten years. During this time the drugs are tested to determine:

➤ that they work
➤ that they are safe
➤ that they are given at the correct **dose** (early tests usually involve low doses).

New drugs made in laboratory → Drugs tested in laboratory for toxicity using cells, tissues and live animals → Clinical trials involving healthy volunteers and patients to check for side-effects

In addition to testing, **computer models** are used to predict how the drug will affect cells, based on knowledge about how the body works and the effects of similar drugs. There are many who believe this type of testing should be extended and that animal testing should be phased out.

Clinical trials

Clinical trials are carried out on healthy volunteers and patients who have the disease. Some are given the new drug and others are given a **placebo**. The effects of the drug can then be compared to the effects of taking the placebo.

Blind trials involve volunteers who do not know if they have been given the new drug or a placebo. This eliminates any psychological factors and helps to provide a fair comparison. (Blind trials are not normally used in modern clinical trials.)

Double blind trials involve volunteers who are randomly allocated to groups. **Neither they nor the doctors/scientists** know if they have been given the new drug or a placebo. This eliminates **all** bias from the test.

New drugs must also be tested against the best existing treatments.

Keywords

Active ingredient ➤ Chemical in a drug that has a therapeutic effect (other chemicals in the drug simply enhance flavour or act as bulking agents)
Pharmaceutical drug ➤ Chemicals that are developed artificially and taken by a patient to relieve symptoms of a disease or treat a condition
Placebo ➤ A substitute for the medication that does not contain the active ingredient
HT Hybridoma cells ➤ Cells produced by the fusion of lymphocytes and a tumour cell

WS When studies involving new drugs are published, there is a **peer review**. This is where scientists with appropriate knowledge read the scientific study and examine the data to see if it is **valid**. Sometimes the trials are duplicated by others to see if similar results are obtained. This increases the **reliability** of the findings and filters out false or exaggerated claims.

Once a **consensus** is agreed, the paper is published. This allows others to hear about the work and to develop it further.

In the case of pharmaceutical drugs, clinical bodies have to decide if the drug can be **licensed** (allowed to be used) and whether it is **cost-effective**. This can be controversial because a potentially life-saving drug may not be used widely simply because it costs too much and/or would benefit too few people.

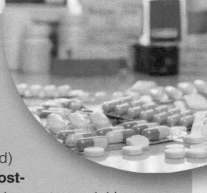

HT Monoclonal antibodies

Monoclonal antibodies are used to treat diseases and in technological applications. They are artificially produced in the laboratory.

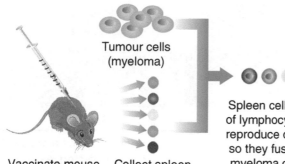

Vaccinate mouse to stimulate the production of antibodies

Collect spleen cells that form antibodies from mouse

Tumour cells (myeloma)

Spleen cells are a type of lymphocyte and can't reproduce on their own, so they fuse (join) with myeloma cells to form hybridoma cells

Grow hybridoma cells in tissue culture and select antibody-forming cells

Collect monoclonal antibodies

The **hybridoma cells** are important because they divide rapidly (by cloning) and produce the required antibody. This means that millions of identical cloned cells are made, which can then be **purified** and used.

Uses of monoclonal antibodies

Use	How?	Advantages
Cancer treatment	Antibodies are made that react to an antigen found on a cancer cell, e.g. in a pancreatic cancer tumour. A toxic agent, such as a radioactive substance, is bound to the antibody and can then be injected into the patient.	The antibody enables the agent to be carried directly to cancer cells and does not harm other cells in the body (unlike chemotherapy and radiotherapy). The technique can also be used to locate blood clots.
Pregnancy testing kits	HCG is a hormone produced in pregnancy. Antibodies to the protein hormone are bound to a test strip and used to detect HCG in urine.	Rapid results (within minutes).
Measuring hormone levels in laboratories	Antibodies can be used to detect hormones or to detect pathogens and other chemicals in the blood.	Provides new information that wouldn't otherwise have been available.
Research	Used to locate a specific molecule in cells or tissues by binding to them with a fluorescent marker.	Marking of molecules made more specific.

When using monoclonal antibodies to treat cancer, more side-effects than expected can arise.

    1. What are the alternatives to testing drugs on animals?

HT **2.** List two uses of monoclonal antibodies.

Plant diseases

Keywords

HT Diagnostic ➤ A process used to identify or recognise a disease or pathogen

Chlorosis ➤ Where leaves lose their colour as a result of mineral deficiency

HT Detecting and identifying plant diseases

Plant diseases and pests cause widespread damage to plants in the wild and in agriculture. They threaten food stocks, affect ecosystems and cause a loss to the economy.

Identifying and detecting plant diseases involves several stages.

| Look for: • stunted growth • leaf spots • areas of decay • growths/tumours • malformed stems and leaves • discolouration • presence of pests. | ➤ | Eliminate environmental causes, e.g. abrasion, poor water supply, climatic factors. | ➤ | Study the distribution of affected plants and refer to garden manuals or websites. | ➤ | Carry out **diagnostic** testing in a laboratory to identify pathogens, e.g. observation; using monoclonal antibody testing kits; DNA testing. |

Pathogens and pests that affect plants

Disease	Pathogen	Appearance/effect on plants	Treatment
Rose black spot	Fungal disease – the fungal spores are spread by water and wind	Purple/black spots on leaves; these then turn yellow and drop early, leading to a lack of photosynthesis and poor growth	Apply a fungicide and/or remove affected leaves
Tobacco mosaic virus (TMV)	Widespread disease that affects many plants (including tomatoes)	'Mosaic' pattern of discolouration; can lead to lack of photosynthesis and poor growth	
Ash dieback	Caused by the fungus *Chalara*	Leaf loss and bark lesions	Cut back or remove diseased trees to reduce chance of airborne spores
Barley powdery mildew	*Erysiphe graminis*	Causes powdery mildew to appear on grasses, including cereals	Fungicides and careful application of nitrogen fertilisers
Crown gall disease	*Agrobacterium tumefaciens*	Tumours or 'galls' at the crown of plants such as apple, raspberry and rose	Use of copper and methods of biological control

Pests	What they do	Appearance/effect on plants	Control
Invertebrates and particularly insects, e.g. many species of aphids	Feed on sap, leaves and storage organs; transmit pathogenic viruses		Chemical pesticides or biological control methods

Mineral ion deficiencies

Plants need **mineral ions** to build complex molecules. The ions are obtained from the soil via the roots in an active manner (requiring energy). In particular, plants need:

➤ **nitrates** to form **amino acids**, the building blocks of **proteins**. They are also needed to make nucleic acids such as **DNA**. Lack of nitrates in a plant leads to yellow leaves and stunted growth

➤ **magnesium** to form chlorophyll, which absorbs light energy for photosynthesis. Lack of magnesium results in **chlorosis**, which is a discolouration of the leaves.

Defence responses of plants

Plants defend themselves from herbivores and disease in a range of ways. These adaptations have evolved over millions of years to maximise a plant's survival.

Mechanical defences include:
➤ thorns and hairs to deter plant-eaters
➤ leaves that droop or curl on contact
➤ mimicry, to fool animals into avoiding them as food or laying their eggs on them. For example, passion flowers have structures that look like yellow eggs. A butterfly is less likely to lay eggs on a plant that has apparently already been used.

Physical defences include:
➤ tough, waxy leaf cuticles
➤ cellulose cell walls
➤ layers of dead cells around stems, e.g. bark. These fall off, taking pathogens with them.

Chemical defences include:
➤ producing antibacterial chemicals, e.g. mint and witch hazel
➤ producing toxins to deter herbivores, e.g. deadly nightshade, foxgloves and tobacco plants.

Revise and test yourself on this topic by designing a set of question/answer cards.
➤ Each card should look like a domino with two halves.
➤ On one side, write a question. On the other, write the answer to a question from the next card. The last card's answer should loop back to the first card's question.
➤ Then shuffle and re-organise the cards and start playing!

1. List three things that plant scientists look for when detecting whether a plant is diseased.
2. Why is the tobacco mosaic virus so damaging to plants?

🎧 18

Mind map

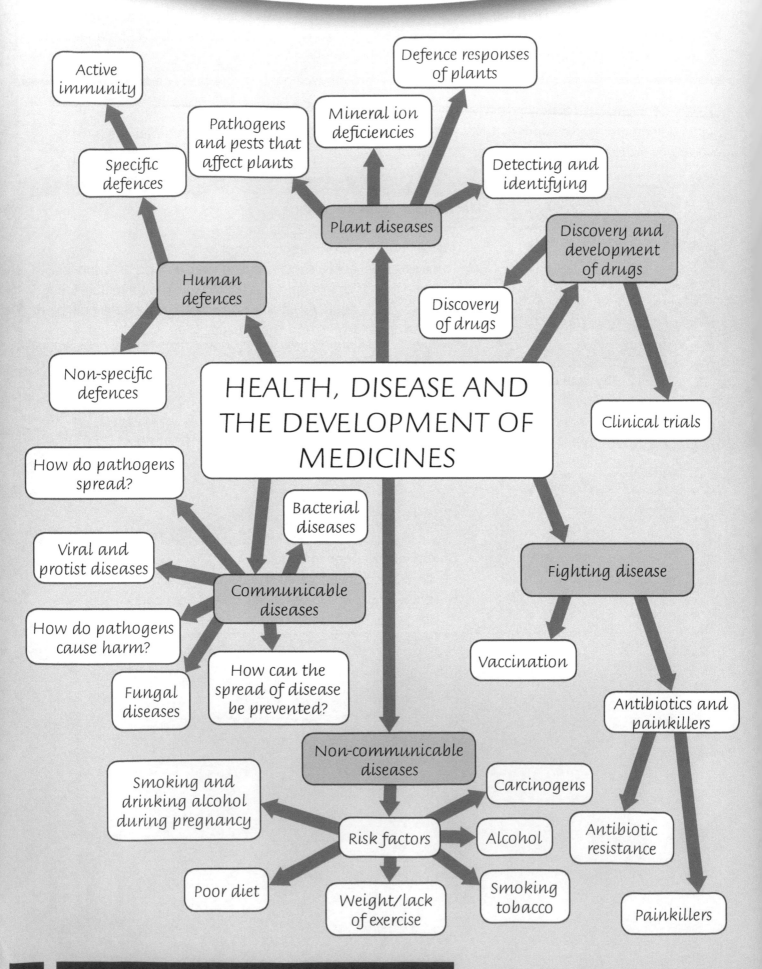

Active immunity

Specific defences

Pathogens and pests that affect plants

Mineral ion deficiencies

Defence responses of plants

Detecting and identifying

Human defences

Plant diseases

Discovery and development of drugs

Discovery of drugs

Non-specific defences

HEALTH, DISEASE AND THE DEVELOPMENT OF MEDICINES

Clinical trials

How do pathogens spread?

Bacterial diseases

Viral and protist diseases

Communicable diseases

Fighting disease

How do pathogens cause harm?

Fungal diseases

How can the spread of disease be prevented?

Vaccination

Antibiotics and painkillers

Non-communicable diseases

Smoking and drinking alcohol during pregnancy

Carcinogens

Risk factors

Alcohol

Antibiotic resistance

Poor diet

Weight/lack of exercise

Smoking tobacco

Painkillers

Practice questions

1. Draw lines to link the name of the microorganism to the disease it causes. **(3 marks)**

Bacterium		HIV
Fungus		Malaria
Virus		Cholera
Protist		Athlete's foot

2. Rani has caught flu and has been confined to bed for several days. Her mother is a health worker and was immunised against flu the previous month.

   a) Describe how the different types of blood component deal with the viruses in Rani's body.

      i) phagocytes **(1 mark)**

      ii) antibodies **(1 mark)**

   b) Describe how the cells in Rani's mother's body responded to the vaccination she was given. In your answer, state what was in the vaccine and use the words **antigen**, **antibody** and **memory cells**. **(4 marks)**

   c) After four days Rani is still unwell and she goes to the doctor. The doctor advises plenty of rest, regular intake of fluids and painkillers when necessary. Explain why the doctor doesn't prescribe antibiotics. **(2 marks)**

   d) If Rani's symptoms continued, which other type of drug might she be prescribed? **(1 mark)**

3. Two drugs called 'Redu' and 'DDD' have been developed to help obese people lose weight. Clinical trials are carried out on the two drugs. The results are shown in the table.

Drug	Number of volunteers in trial	Average weight loss in 6 weeks (kg)
Redu	3250	3.2
DDD	700	5.8
Placebo	2800	2.6

   a) The scientific team concluded that DDD was a more effective weight loss drug. Do you agree? Use the data in the table to give a reason for your answer. **(2 marks)**

   b) The trial carried out was 'double blind'. Explain what this term means and why it is used. **(2 marks)**

Homeostasis and negative feedback

Keywords

Endocrine system ➤ System of ductless organs that release hormones

Stimuli ➤ Changes in the internal or external environment that affect receptors

Homeostasis

The body has automatic control systems to maintain a constant internal environment (**homeostasis**). These systems make sure that cells function efficiently.

Homeostasis balances inputs and outputs to ensure that optimal levels of temperature, pH, water, oxygen and carbon dioxide are maintained. For example, even in the cold, homeostasis ensures that body temperature is regulated at about 37°C.

Control systems in the body may involve the nervous system, the **endocrine system**, or both. There are three components of control.

➤ **Effectors** cause responses that restore optimum levels, e.g. muscles and glands.

➤ **Coordination centres** receive and process information from the receptors, e.g. brain, spinal cord and pancreas.

➤ **Receptors** detect **stimuli** from the environment, e.g. taste buds, nasal receptors, the inner ear, touch receptors and receptors on retina cells.

(WS) When taking measurements, the quality of the measuring instrument and a scientist's skill is very important to achieve **accuracy**, **precision** and **minimal error**.

Adrenaline levels in blood plasma are measured by a chromatography method called HPLC. This is often coupled to a detector that gives a digital readout.

This digital readout displays the concentration of adrenaline as 6.32. This means that the instrument is precise up to $\frac{1}{100}$ of a unit.

A less precise instrument might only measure down to $\frac{1}{10}$ of a unit, e.g. 6.3 (one decimal place).

A bar graph of some data generated from HPLC is shown below. The graph shows the **average concentration** of three different samples of blood. The average is taken from many individual measurements. The vertical error bars indicate the range of measurements (the difference between highest and lowest) obtained for each sample.

Sample C shows **the greatest precision** as the individual readings do not vary as much as the others. There is less error so we can be more confident that the average is closer to the **true value** and therefore more **accurate**.

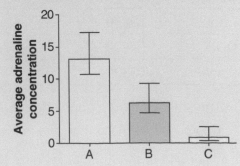

Negative feedback

Negative feedback occurs frequently in homeostasis. It involves the automatic reversal of a change in the body's condition.

In the body, examples of negative feedback include osmoregulation/water balance, balancing blood sugar levels, maintaining a constant body temperature and controlling metabolic rate.

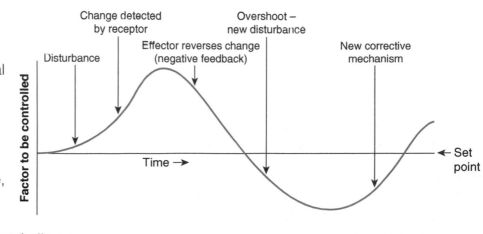

Metabolism needs to be controlled so that chemical reactions in the body take place at an optimal rate. Negative feedback controls metabolic rate by using the hormones **thyroxine** and **adrenaline**.

Thyroxine

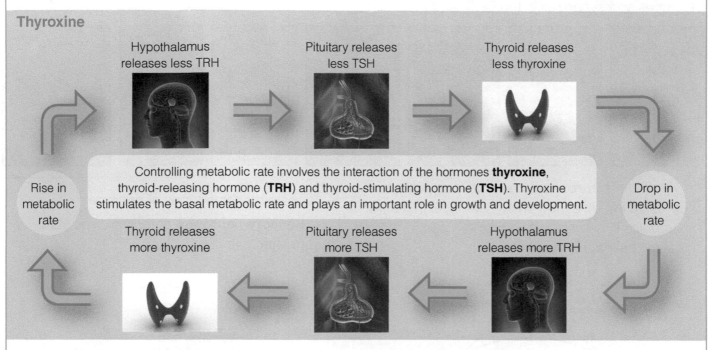

Hypothalamus releases less TRH → Pituitary releases less TSH → Thyroid releases less thyroxine

Controlling metabolic rate involves the interaction of the hormones **thyroxine**, thyroid-releasing hormone (**TRH**) and thyroid-stimulating hormone (**TSH**). Thyroxine stimulates the basal metabolic rate and plays an important role in growth and development.

Rise in metabolic rate

Drop in metabolic rate

Thyroid releases more thyroxine ← Pituitary releases more TSH ← Hypothalamus releases more TRH

Adrenaline

Adrenaline is sometimes called the 'flight or fight' hormone. During times of stress the adrenal glands produce adrenaline. It has a direct effect on muscles, the liver, intestines and many other organs to prepare the body for sudden bursts of energy. Specifically, adrenaline increases the heart rate so that the brain and muscles receive oxygen and glucose more rapidly.

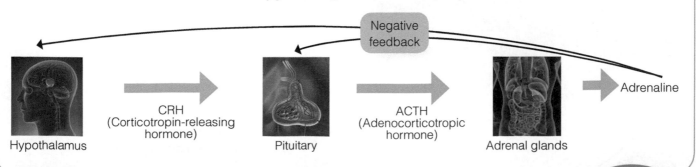

Negative feedback

Hypothalamus — CRH (Corticotropin-releasing hormone) → Pituitary — ACTH (Adenocorticotropic hormone) → Adrenal glands → Adrenaline

HT 1. Give two examples of negative feedback in the human body.
HT 2. What is the target organ for the hormone CRH?

The nervous system

Structure and function

The nervous system allows organisms to react to their surroundings and coordinate their behaviour.

The two main parts of the nervous system are:

➤ the central nervous system (**CNS**), which is made up of the spinal cord and **brain**

➤ the **peripheral nervous system**.

The flow of **impulses** in the nervous system is carried out by nerve cells linking the receptor, coordinator (neurones and synapses in the CNS) and effector.

The main components of the nervous system

Brain

Spinal cord

The neurones that make up the peripheral nervous system

CNS (brain and spinal cord)

Sense organ	Sensory neurone	Synapse	Relay neurone	Synapse	Motor neurone	Muscle
In the sense organ, receptors detect a change – either inside or outside the body. The change is a stimulus.	Conducts the impulse from the sense organ towards the CNS.	The gap between the sensory and relay neurones.	Passes the impulse on to a motor neurone.	The gap between the relay neurone and the motor neurone.	Passes the impulse on to the muscle (or gland).	The muscle responds by contracting, which results in a movement. Muscles and glands are examples of effectors.

Nerve cells or **neurones** are specially adapted to carry nerve impulses, which are electrical in nature. The impulse is carried in the long, thin part of the cell called the **axon**.

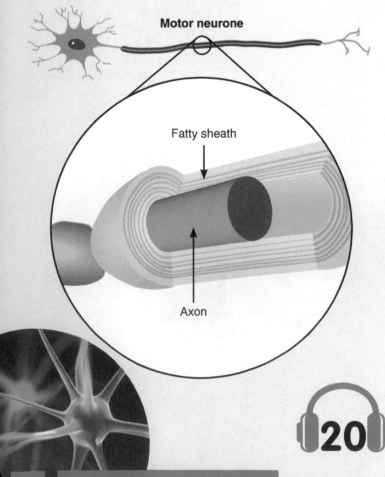

Motor neurone

Fatty sheath

Axon

There are three types of neurone.

Sensory neurones carry impulses from receptors to the CNS.

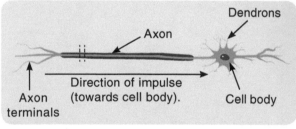

Dendrons

Axon

Direction of impulse (towards cell body).

Axon terminals

Cell body

Relay neurones make connections between neurones inside the CNS.

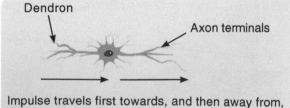

Dendron

Axon terminals

Impulse travels first towards, and then away from, cell body.

Motor neurones carry impulses from the CNS to muscles and glands.

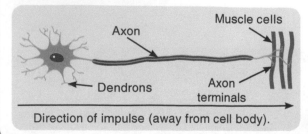

Muscle cells

Axon

Dendrons

Axon terminals

Direction of impulse (away from cell body).

🎧 20

Synapses

Synapses are junctions between neurones. They play an important part in regulating the way impulses are transmitted. Synapses can be found between different neurones, neurones and muscles, and between dendrites (the root-like outgrowths from the cell body).

When an impulse reaches a synapse, a **neurotransmitter** is released by the neurone ('A' in the diagram) into the gap that lies between the neurones. It travels by diffusion and binds to **receptor molecules** on the next neurone. This triggers a new electrical impulse to be released.

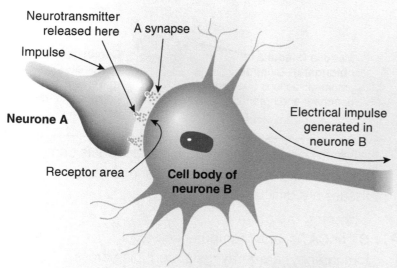

Keywords

Spinal cord ➤ Nervous tissue running down the centre of the vertebral column; millions of nerves branch out from it

Impulses ➤ Electrical signals sent down neurones that trigger responses in the nervous system

Receptor molecule ➤ Protein on the outer membrane of a cell that binds to a specific molecule, such as transmitter substance

Reflex arcs

Reflex actions:

➤ are involuntary/automatic
➤ are very rapid
➤ protect the body from harm
➤ bypass conscious thought.

The pathway taken by impulses around the body is called a **reflex arc**. Examples include:

➤ opening and closing the pupil in the eye
➤ the knee-jerk response
➤ withdrawing your hand from a hot plate.

Here is the arc pathway for a pain response.

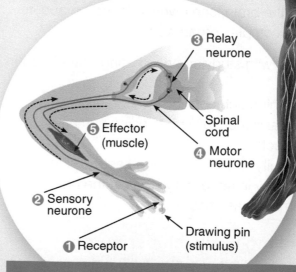

WS You may have to investigate the effect of factors on human reaction time.

For example, you could be asked to investigate a learned reflex by measuring how far up a ruler someone can catch it. The nearer to the zero the ruler is caught, the faster the reflex.

You could investigate factors such as:

➤ experience/practice at catching
➤ sound ➤ touch ➤ sight.

Can you design experiments to test these variables? Which factors will need to be kept the same?

1. Name the long, thin extensions of nerve cells.
2. How do reflex arcs aid survival of an organism?
3. Design and draw a flow diagram to represent the pathway followed by impulses in the knee-jerk reflex. (The flow diagram at the beginning of this module should help you.)

Nervous control

Keywords

Neuroscience ➤ Study of the nervous system

Dermis ➤ The layer of tissue in the skin that lies beneath the epidermis. It contains many receptors and structures involved in temperature control

Epidermis ➤ Outer layer of tissue in the skin

The brain

The brain is part of the central nervous system (**CNS**) and is studied by **neuroscientists**. The brain controls many activities in the body but is also responsible for memory, learning and behaviour.

The brain is made up of billions of neurones, each interlinked via synapses. There are four main parts of the brain.

Frontal lobe – controls higher mental functions, e.g. choice and memory

Cerebral cortex – responsible for numerical computation, language and emotions

Medulla (medulla oblongata) – controls automatic actions such as heartbeat and breathing

Cerebellum – coordinates movement and balance (via muscle activity)

HT Studying the brain

Neuroscientists have mapped the regions of the brain to different functions. This is particularly useful for studying people suffering from brain damage and brain disorders, such as Alzheimer's disease. Studying healthy volunteers, using MRI technology and electrical stimulation of different parts of the brain, can reveal information about how this complex organ works.

Problems with studying and treating brain function include:

➤ obtaining subjects to study
➤ ethical issues relating to using human subjects
➤ difficulty when interpreting case studies
➤ difficulty in accessing areas of the brain where damage has occurred.

When treating nerve damage, the prospects can be limited because nervous and surrounding tissues are difficult to repair.

Scanning the brain

CT or **CAT** scans (computerised axial tomography) are X-ray tests producing cross-sectional images of the brain using computer programs. Many images can be taken at different angles to create 3D pictures.

PET stands for positron emission tomography. A PET scan can show what body tissues look like and how they are working. PET scanners work by detecting radiation given off by radiotracers, which the patient takes orally. The tracers accumulate in particular areas of the body. PET is often combined with CT and MRI technology to give 3D views of the brain.

Controlling body temperature

The thermoregulatory centre in the brain monitors and controls body temperature.

As you found out in Module 6, enzymes work at around **37°C** (core temperature for humans). It is essential that this temperature does not fluctuate too much. Body heat is released by chemical reactions in the body (metabolism) and distributed via the bloodstream.

The skin plays a major role in temperature regulation. The structures responsible are found in the **dermis** and **epidermis** of the skin.

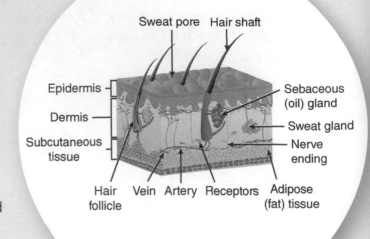

Sweat pore   Hair shaft

Epidermis

Dermis

Subcutaneous tissue

Sebaceous (oil) gland

Sweat gland

Nerve ending

Hair follicle   Vein   Artery   Receptors   Adipose (fat) tissue

When core temperature falls

➤ There is reduced blood flow near the skin to limit heat loss to the environment. This is called **vasoconstriction**.
➤ Tiny involuntary contractions of muscles (shivering) generate heat.
➤ Sweating reduces.
➤ Body hairs rise due to the contraction of special muscles. Humans have less body hair than most mammals so there is minimal benefit.

When core temperature rises

➤ Blood flows closer to the skin surface because blood vessels widen. This is called **vasodilation**.
➤ Sweating increases. The sweat evaporates on the skin's surface, which absorbs heat from the skin in an **endothermic** change.
➤ Body hairs lower.

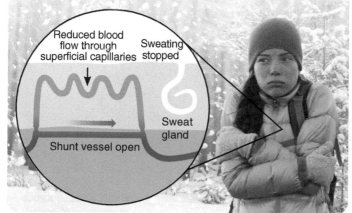

Negative feedback system

There is a negative feedback system for regulating body temperature.

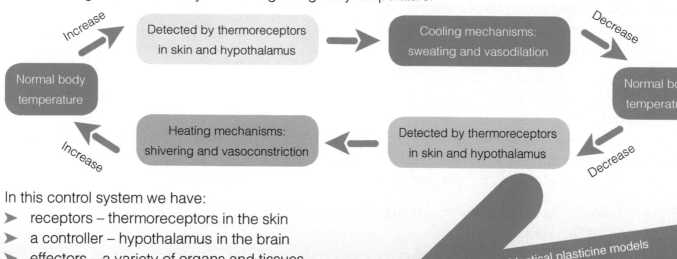

In this control system we have:

➤ receptors – thermoreceptors in the skin
➤ a controller – hypothalamus in the brain
➤ effectors – a variety of organs and tissues.

If a control system breaks down it can be life-threatening.

➤ **Hypothermia** is when the body temperature falls well below 37°C. Symptoms include shivering, blue skin colour, disorientation and unconsciousness.
➤ **Heat stroke/dehydration** is when the body temperature rises well above 37°C. Symptoms include an altered mental state, nausea, flushed skin, and rapid breathing and pulse.

➤ Create two identical plasticine models showing the main parts of the skin involved in temperature control. Use different colours for each of the structures.
➤ Modify one of your models to show what happens if body temperature is rising, and the other one to show what happens when body temperature falls below normal. In particular, try to show what happens to the blood vessels during vasodilation and vasoconstriction.

1. Which part of the brain is responsible for learning, thinking and memory?
HT 2. State the main differences between a CT scan and a PET scan.
3. Describe how structures in the skin respond when core body temperature rises.

The eye

Structure

The eye is a major sense organ in the human body. The receptors are on the **retina** – they consist of light-sensitive cells that detect the intensity and colour of light that enters the eye.

Light-sensitive cells are of two types – rods and cones. Rods are sensitive to low light intensities and enable vision in black and white. Cones are less sensitive and operate at higher light intensities. They enable us to see in colour. The three types of cone are those sensitive to:

➤ red light
➤ green light
➤ blue light.

Light entering the eye is perceived as a certain colour because these three types of cone are stimulated to different degrees, resulting in many millions of different shades and hues.

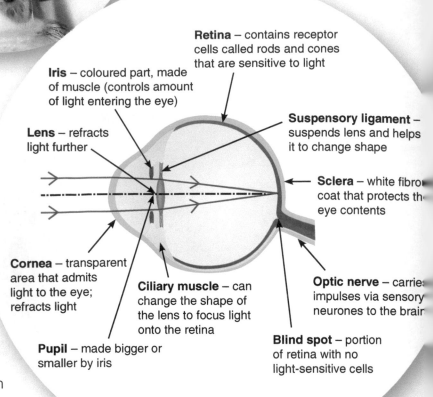

Retina – contains receptor cells called rods and cones that are sensitive to light

Iris – coloured part, made of muscle (controls amount of light entering the eye)

Suspensory ligament – suspends lens and helps it to change shape

Lens – refracts light further

Sclera – white fibrous coat that protects the eye contents

Cornea – transparent area that admits light to the eye; refracts light

Ciliary muscle – can change the shape of the lens to focus light onto the retina

Optic nerve – carries impulses via sensory neurones to the brain

Pupil – made bigger or smaller by iris

Blind spot – portion of retina with no light-sensitive cells

Accommodation

The function of the eye is to receive light rays and focus them to form a sharp image on the retina. The optical information is sent to the brain, which interprets it to give a sense of a 3D image, movement, colour and brightness. This interpretation by the occipital lobe in the brain is called **perception**.

The focusing process is called **accommodation**. It involves changing the shape of the lens in order to refract light.

To focus on a **near object**: ➤ light rays **diverge** significantly ➤ the ciliary muscles **contract** ➤ the suspensory ligaments loosen/go **slack** ➤ the lens becomes **short** and **fat** (**more** convex) ➤ the direction of light rays changes **greatly** to converge on the retina.	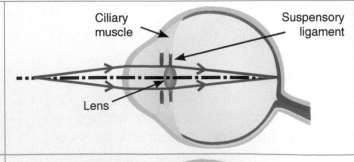
To focus on a distant object: ➤ light rays are almost parallel ➤ the ciliary muscles **relax** ➤ the suspensory ligaments become **taut** ➤ the lens becomes **long** and **thin** (**less convex**) ➤ the direction of light rays changes **slightly** to converge on the retina.	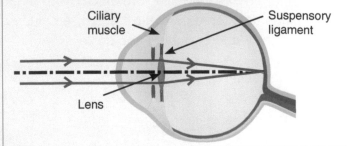

This is the front view of what happens during accommodation.

Near object

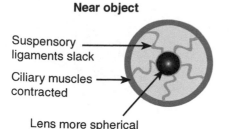

Suspensory ligaments slack

Ciliary muscles contracted

Lens more spherical

Distant object

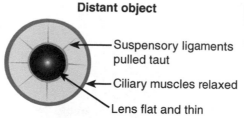

Suspensory ligaments pulled taut

Ciliary muscles relaxed

Lens flat and thin

Keywords

Diverge ➤ Spread outwards from a point

Convex ➤ Lens that curves outwards on one or both sides

Converge ➤ Come together at a point

Concave ➤ Lens that curves inwards on one or both sides

Defects

Common eye defects are:

➤ hyperopia (long sightedness)
➤ myopia (short sightedness)
➤ red–green colour blindness
➤ cataracts.

Cataract

Hyperopia and **myopia** can be corrected using spectacles, contact lenses or laser surgery (to change the shape of the cornea).

Hyperopia is:

➤ caused by an eyeball that is too short, or a lens that stays too long and thin
➤ corrected by a **convex** lens.

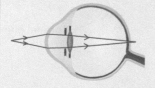 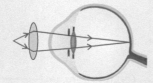

Myopia is:

➤ caused by an eyeball that is too long, or weak suspensory ligaments that cannot pull the lens into a thin shape
➤ corrected by a **concave** lens.

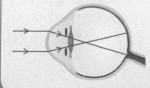 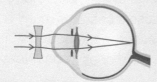

Red–green colour blindness is more common in boys as it is determined by a defective gene on the X chromosome. The result is one or more cone types being faulty, causing an inability to distinguish between different shades of red or green.

A **cataract** develops when the lens clouds over. This is caused by some of the protein in the lens clumping together. Using spectacles with strong bifocals is a temporary treatment. The only permanent solution is removal during surgery.

An alternative to spectacles is **contact lenses**. These are miniature lenses (either hard or soft) that can be placed over the front of the eye. They need to be kept clean if they are a permanent type, or disposed of if temporary. Some operations can replace the natural lens with a new one.

WS You will be expected to give examples of how scientific ideas can lead to technical applications. The development of optical instruments for helping vision is a good example.

Sir Isaac Newton developed laws of optics in the seventeenth century. This built on theories put forward by scientists previously. An understanding of how light is refracted and reflected, and of ray diagrams, has led to inventions such as telescopes, microscopes and spectacles.

22

1. Which part of the eye controls the amount of light entering it?
2. Describe the shape of the lens when the eye is focusing on an object a short distance away.
3. What causes hyperopia?

The endocrine system

Keywords

Ductless gland ➤ A gland that does not secrete its chemicals through a tube. The pancreas is an exception to this rule as it contains a duct for delivering enzymes, but its hormones are released directly into the bloodstream

Glucagon ➤ Hormone released by the pancreas that stimulates the conversion of glycogen to glucose

Glycogen ➤ Storage carbohydrate found in animals

Structure and function

The endocrine system is made up of glands that are **ductless** and secrete **hormones** directly into the bloodstream. The blood carries these chemical messengers to **target organs** around the body, where they cause an effect.

Hormones:

➤ are large, protein molecules

➤ interact with the nervous system to exert control over essential biological processes

➤ act over a longer time period than nervous responses but their effects are slower to establish.

Endocrine gland	Hormone(s) produced
Pituitary gland	TSH, ADH, FSH, LH, etc.
Pancreas	Insulin, **glucagon**
Thyroid	Thyroxine
Adrenal gland	Adrenaline
Ovaries (female)	Oestrogen, progesterone
Testes (male)	Testosterone

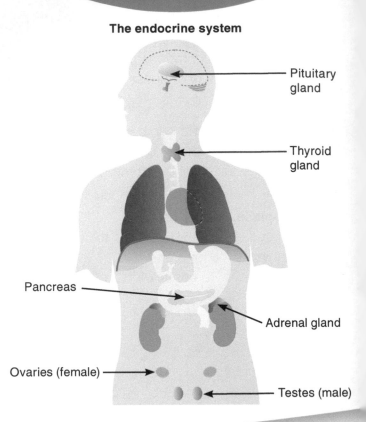

The endocrine system

- Pituitary gland
- Thyroid gland
- Pancreas
- Adrenal gland
- Ovaries (female)
- Testes (male)

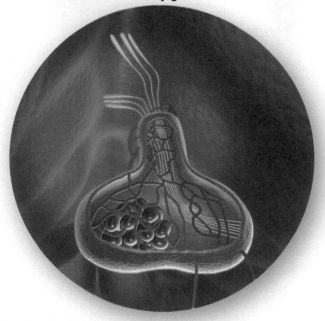

Pituitary gland

The pituitary gland

The pituitary is often referred to as the **master gland** because it secretes many hormones that control other processes in the body. Pituitary hormones often trigger other hormones to be released.

➤ Create a series of flashcards about controlling blood glucose levels. On each one, write a hormone, organ or effect relating to the control system. Shuffle the cards and put them face down.

➤ With a revision buddy, or on your own, pick up each card and explain how the particular component is involved in the control process.

Controlling blood glucose concentration

The control system for balancing blood glucose levels involves the **pancreas**.

The pancreas monitors the blood glucose concentration and releases hormones to restore the balance. When the concentration is too high, the pancreas produces insulin that causes glucose to be absorbed from the blood by all body cells, but particularly those in the liver and muscles. These organs convert glucose to **glycogen** for storage until required.

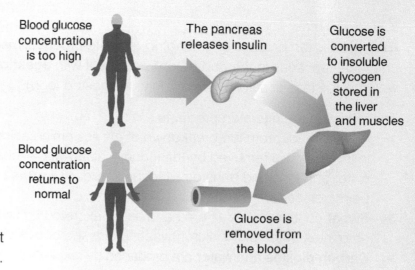

Blood glucose concentration is too high → The pancreas releases insulin → Glucose is converted to insoluble glycogen stored in the liver and muscles → Glucose is removed from the blood → Blood glucose concentration returns to normal

HT If blood glucose concentration is too low, the pancreas secretes glucagon. This stimulates the conversion of glycogen to glucose via enzymic systems. It is then released into the blood.

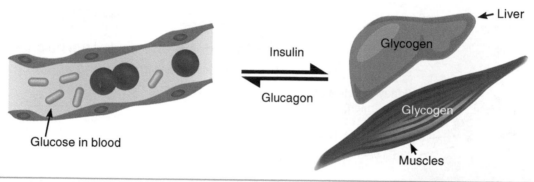

Glucose in blood — Insulin / Glucagon — Liver (Glycogen), Muscles (Glycogen)

Diabetes
There are two types of diabetes.

Type I diabetes:
- is caused by the pancreas' inability to produce insulin
- results in dangerously high levels of blood glucose
- is controlled by delivery of insulin into the bloodstream via injection or a 'patch' worn on the skin
- is more likely to occur in people under 40
- is the most common type of diabetes in childhood
- is thought to be triggered by an auto-immune response where cells in the pancreas are destroyed.

Type II diabetes:
- is caused by fatty deposits preventing body cells from absorbing insulin; the pancreas tries to compensate by producing more and more insulin until it is unable to produce any more
- results in dangerously high levels of blood glucose
- is controlled by a low carbohydrate diet and exercise initially; it may require insulin in the later stages
- is more common in people over 40
- is a risk factor if you are obese.

1. Why is the pituitary called the master gland?
2. Where are the sex hormones of the body produced?
3. What effects does type II diabetes have on the body?

Excretion

Excretion is the process of getting rid of waste products made by chemical reactions in the body. Don't confuse it with **egestion**, which is the loss of solid waste (mainly undigested food).

The following are excreted products.

➤ **Urea** is made from the breakdown of excess amino acids in the liver. It is removed by the kidneys along with excess water and ions and transferred to the bladder as **urine** before being released.

➤ **Sweat** containing water, urea and salt is excreted by sweat glands onto the surface of the skin. Sweating aids the body's cooling process.

➤ **Carbon dioxide** and **water** are produced by respiration and leave the body from the lungs during exhalation.

The lungs and skin don't control the loss of substances. They are simply the organs by which these substances are removed.

Water & nitrogen balance

The kidneys

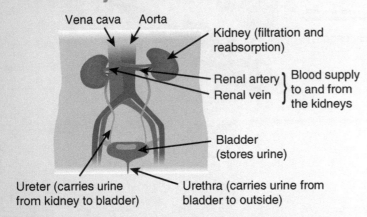

Vena cava   Aorta

Kidney (filtration and reabsorption)

Renal artery } Blood supply
Renal vein } to and from the kidneys

Bladder (stores urine)

Ureter (carries urine from kidney to bladder)

Urethra (carries urine from bladder to outside)

The kidneys filter the blood, allowing urea to pass to the bladder. The filtering is carried out within the kidney by thousands of tiny **kidney tubules**.

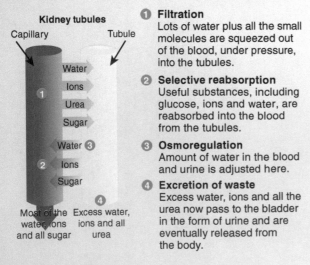

Kidney tubules

Capillary   Tubule

Water
Ions
Urea
Sugar

Water ❸
Ions
Sugar

❹

Most of the water, ions and all sugar   Excess water, ions and all urea

❶ **Filtration**
Lots of water plus all the small molecules are squeezed out of the blood, under pressure, into the tubules.

❷ **Selective reabsorption**
Useful substances, including glucose, ions and water, are reabsorbed into the blood from the tubules.

❸ **Osmoregulation**
Amount of water in the blood and urine is adjusted here.

❹ **Excretion of waste**
Excess water, ions and all the urea now pass to the bladder in the form of urine and are eventually released from the body.

Other useful substances (glucose, amino acids, fatty acids, glycerol and some water) are **selectively re-absorbed** early in the process.

The kidneys also control the balance of water in the blood. Damage occurs to red blood cells if water content is not balanced.

Ideal shape	Swollen	Shrivelled
When red blood cells (erythrocytes) are in solutions with equal concentration to their cytoplasm, they have an ideal, biconcave shape. This is because there is no net movement of water in or out.	When immersed in a solution of lower concentration (higher water concentration), the cells absorb water by osmosis. The weak cell membrane cannot resist the added water pressure and may burst.	In a more concentrated solution (lower water concentration), cells lose water by osmosis. They shrivel up and become **crenated** (have scalloped edges).

Kidney tubules (nephrons)

In terms of nephron structure, filtration, selective reabsorption and excretion of waste occur in the following regions.

Structure of the nephron

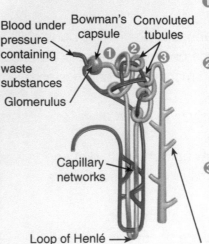

Blood under pressure containing waste substances

Bowman's capsule   Convoluted tubules

Glomerulus

Capillary networks

Loop of Henlé

Collecting ducts (lead to the ureter)

❶ Filtration, where all small molecules and lots of water are squeezed out of the blood and into the tubules.

❷ Selective reabsorption of useful substances (water, ions, glucose) back into the blood from the convoluted tubules. This may take energy in the case of glucose and ions.

❸ Excretion of waste in the form of excess water, excess ions and all urea. These drain into the collecting tubules and pass to the bladder as urine.

Kidney failure

Kidneys may fail due to accidents or disease. A patient can survive with one kidney. If both kidneys are affected, two treatments are available.

> **Kidney transplant** – involves a healthy person donating one kidney to replace two failed kidneys in another person.

> **Dialysis** – offered to patients while they wait for the possibility of a kidney transplant. A dialysis machine removes urea and maintains levels of sodium and glucose in the blood.

This is what happens during dialysis.

1. Blood is taken from a person's vein and run into the dialysis machine, where it comes into close contact with a **partially permeable membrane**.
2. This separates the blood from the dialysis fluid.
3. The urea and other waste diffuse from the blood into the dialysis fluid. The useful substances remain and are transferred back to the body.

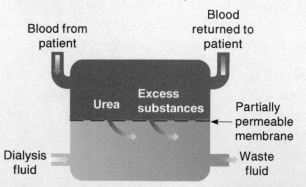

24

Keywords

Urea ➤ Nitrogenous waste product

Selective re-absorption ➤ Occurs in the kidney tubules and allows useful substances to pass from the kidneys back into the blood

Partially permeable membrane ➤ Artificial or organic layer that only allows small molecules through

HT Deamination ➤ Process in the liver in which nitrogen is removed from an amino acid molecule

HT Ammonia ➤ Nitrogenous waste product

HT How urea is formed

Proteins obtained from the diet may produce a surplus of **amino acids** that need to be excreted safely.

1. First, the amino acids are **deaminated** in the **liver** to form **ammonia**.
2. Ammonia is toxic so is immediately converted to urea, which is then filtered out in the kidney.

Controlling water content

The osmotic balance of the body's fluids needs to be tightly controlled because if cells gain or lose too much water they do not function efficiently.

The amount of water re-absorbed by the kidneys is controlled by **anti-diuretic hormone (ADH)**. This is produced in the pituitary.

1. ADH directly increases the permeability of the kidney tubules to water.
2. When the water content of the blood is low (higher blood concentration), **negative feedback** operates to restore normal levels.

The effect of ADH on blood water content

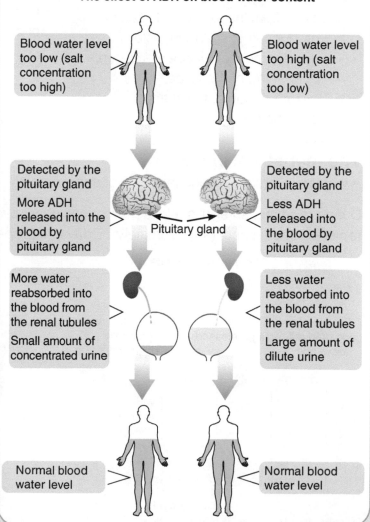

---

1. List three substances that are selectively re-absorbed back into the bloodstream from the kidney.
2. **HT** What effect does producing **more** ADH have on the concentration of urine?
3. Describe what happens to a urea molecule as it passes through a dialysis machine.

Hormones in human reproduction

Hormones play a vital role in regulating human reproduction, especially in the female **menstrual** cycle.

Puberty

During **puberty** (approximately 10–16 in girls and 12–17 in boys), the sex organs begin to produce **sex hormones**. This causes the development of **secondary sexual characteristics**.

In **males**, the primary sex hormone is **testosterone**.

During puberty, testosterone is produced from the testes and causes:

➤ production of sperm in testes
➤ development of muscles and penis
➤ deepening of the voice
➤ growth of pubic, facial and body hair.

In **females**, the primary sex hormone is **oestrogen**. Other sex hormones are **progesterone**, **FSH** and **LH**.

During puberty, oestrogen is produced in the ovaries and progesterone production starts when the menstrual cycle begins.

The secondary sexual characteristics are:

➤ ovulation and the menstrual cycle
➤ breast growth
➤ widening of hips
➤ growth of pubic and armpit hair.

The menstrual cycle

A woman is fertile between the ages of approximately 13 and 50.

During this time, an egg is released from one of her ovaries each month and the lining of her uterus is replaced each month (approximately 28 days) to prepare for pregnancy.

The menstrual cycle

Follicle with egg gradually develops

Ovulation (egg released)

Empty follicle gradually disappears

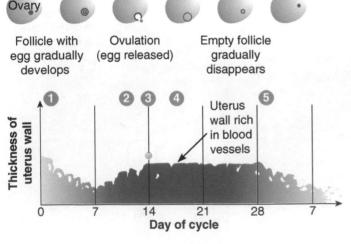

Thickness of uterus wall

Uterus wall rich in blood vessels

Day of cycle

① Uterus lining breaks down (i.e. a period).
② Repair of the uterus wall. Oestrogen causes the uterus lining to gradually thicken.
③ Egg released by the ovary.
④ Progesterone and oestrogen make the lining stay thick, waiting for a fertilised egg.
⑤ No fertilised egg so cycle restarts.

Concentration of hormones in the blood

Progesterone

Oestrogen

As well as oestrogen and progesterone, the two other hormones involved in the cycle are:

➤ **FSH** or **follicle stimulating hormone**, which causes maturation of an egg in the ovary
➤ **LH** or **luteinising hormone**, which stimulates release of an egg.

Negative feedback in the menstrual cycle

The four female hormones interact in a complex manner to regulate the cycle.

➤ **FSH** is produced in the pituitary and acts on the ovaries, causing an egg to mature. It **stimulates** the ovaries to produce oestrogen.
➤ **Oestrogen** is secreted in the ovaries and **inhibits** further production of FSH. It also **stimulates** the release of LH and promotes repair of the uterus wall after menstruation.
➤ **LH** is produced in the pituitary. It also **stimulates** release of an egg.
➤ **Progesterone** is secreted by the empty **follicle** in the ovary (left by the egg). It **maintains** the lining of the uterus after ovulation has occurred. It also **HT** **inhibits** FSH and LH.

Female reproductive system

Low progesterone levels allow FSH from the pituitary gland to stimulate the maturation of an egg (in a follicle). This in turn stimulates oestrogen production.

High levels of oestrogen stimulate a surge in LH from the pituitary gland. This triggers ovulation in the middle of the cycle.

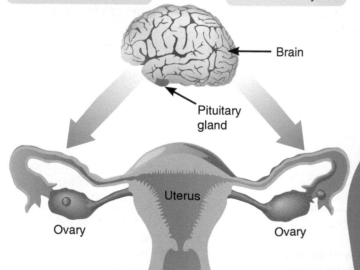

Brain

Pituitary gland

Uterus

Ovary   Ovary

Keywords

Menstruation ➤ Loss of blood and muscle tissue from the uterus wall during the monthly cycle

Follicle stimulating hormone ➤ A hormone produced by the pituitary gland that controls oestrogen production by the ovaries

HT Inhibition ➤ Negative effect in negative feedback, where the increase in one factor brings a **decrease** in another

25

To familiarise yourself with what the various hormones do in the menstrual cycle, you may find it helpful to do the following.

➤ On a computer, create a series of text boxes listing the organs, hormones and effects relating to the menstrual cycle. You could also include diagrams of the organs.
➤ Re-arrange the text boxes/diagrams so they are out of order.
➤ Test yourself or ask a revision buddy to put them back in the correct position/order.

Alternatively, you could draw or write the organs, hormones and effects on paper and re-arrange them manually.

1. Name the four hormones involved in the menstrual cycle. What are their functions?
2. At approximately what stage in the menstrual cycle does the uterus lining repair itself?
3. Name one effect of testosterone in puberty.

Contraception and infertility

Fertility and the possibility of pregnancy can be controlled using non-hormonal and hormonal methods of **contraception**.

Keywords

Contraception ➤ Literally means 'against conception'; any method that reduces the likelihood of a sperm meeting an egg
Intrauterine ➤ Inside the uterus
Implantation ➤ Process in which an embryo embeds itself in the uterine wall
Oral contraceptive ➤ Hormonal contraceptive taken in tablet form

Non-hormonal contraception

Contraceptive method	Method of action	Advantages	Disadvantages
➤ Barrier method – condom (male + female)	Prevents the sperm from reaching the egg	82% effective ➤ Most effective against STIs	➤ Can only be used once ➤ May interrupt sexual activity ➤ Can break ➤ Women may be allergic to latex
➤ Barrier method – diaphragm	Prevents the sperm from reaching the egg	88% effective ➤ Can be put in place right before intercourse or 2–3 hours before ➤ Don't need to take out between acts of sexual intercourse	➤ Increases urinary tract infections ➤ Doesn't protect against STIs
➤ **Intrauterine** device	Prevents **implantation** – some release hormones	99% effective ➤ Very effective against pregnancy ➤ Doesn't need daily attention ➤ Comfortable ➤ Can be removed at any time	➤ Doesn't protect against STIs ➤ Needs to be inserted by a medical practitioner ➤ Higher risk of infection when first inserted ➤ Can have side effects such as menstrual cramping ➤ Can fall out and puncture the uterus (rare)
➤ Spermicidal agent	Kills or disables sperm	72% effective ➤ Cheap	➤ Doesn't protect against STIs ➤ Needs to be reapplied after one hour ➤ Increases urinary tract infections ➤ Some people are allergic to spermicidal agents
➤ Abstinence ➤ Calendar method	Refraining from sexual intercourse when an egg is likely to be in the oviduct	76% effective ➤ Natural ➤ Approved by many religions ➤ Woman gets to know her body and menstrual cycles	➤ Doesn't protect against STIs ➤ Calculating the ovulation period each month requires careful monitoring and instruction ➤ Can't have sexual intercourse for at least a week each month
➤ Surgical method	Vasectomy and female sterilisation	99% effective ➤ Very effective against pregnancy ➤ One-time decision providing permanent protection	➤ No protection against STIs ➤ Need to have minor surgery ➤ Permanent

Hormonal contraception

Contraceptive method	Method of action	Advantages	Disadvantages
➤ Oral contraceptive 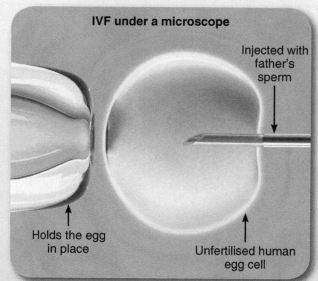	Contains hormones that inhibit FSH production, so eggs fail to mature	91% effectiveness ➤ Very effective against pregnancy if used correctly ➤ Makes menstrual periods lighter and more regular ➤ Lowers risk of ovarian and uterine cancer, and other conditions ➤ Doesn't interrupt sexual activity	➤ Doesn't protect against STIs ➤ Need to remember to take it every day at the same time ➤ Can't be used by women with certain medical problems or by women taking certain medications ➤ Can occasionally cause side effects
➤ Hormone injection ➤ Skin patch ➤ Implant	Provides slow release of progesterone; this inhibits maturation and release of eggs	91–99% effectiveness depending on method used ➤ Lasts over many months or years ➤ Light or no menstrual periods ➤ Doesn't interrupt sexual activity	➤ Doesn't protect against STIs ➤ May require minor surgery (for implant) ➤ Can cause side effects

The percentage figures in the contraception tables are based on users in a whole population, regardless of whether they use the method correctly. If consistently used correctly, the percentage effectiveness of each method is usually higher. Some methods, such as the calendar method, are more prone to error than others.

HT Infertility treatment

Infertility treatment is used by couples who have problems conceiving.

Reasons for infertility

No eggs being released from the ovaries.
Endometriosis, which occurs when the tissue that lines the inside of the uterus enters other organs of the body, such as the abdomen and fallopian tubes; this reduces the maturation rate and release of eggs.
Male infertility/low sperm count.
Uterine fibroids.
Complete or partial blocking and/or scarring of the fallopian tubes.
Reduced number and quality of eggs.

Methods of treatment

Treating infertility is known as ART (assisted reproductive technology). There are a number of methods of treatment.

➤ **Fertility drugs** containing FSH and LH are given to women who do not produce enough FSH themselves. They may then become pregnant naturally.

➤ **Clomifene therapy** prevents the production of oestrogen and so inhibits negative feedback.

➤ **In vitro fertilisation (IVF)** is a method in which the potential mother is given FSH and

LH to stimulate the production of several eggs. Sperm is collected from the father. The sperm and eggs are then introduced together outside the body in a petri dish. One or two growing embryos can then be transplanted into the woman's uterus.

IVF under a microscope

Injected with father's sperm

Holds the egg in place

Unfertilised human egg cell

Disadvantages of IVF

It is very expensive.
It can be mentally and physically stressful.
Success rates are only approximately 40% (at the time of writing).
There is an increased risk of multiple births.

1. Which contraceptives might not be suitable for a woman suffering from high blood pressure?
2. Why are condoms effective against the spread of HIV?
HT 3. How has the development of microscopy helped couples with infertility problems?

Plant hormones

Tropisms and general control

Plants, as well as animals, respond to changes in their environment.

Plant hormones are chemicals that control the growth of shoots and roots, flowering and the ripening of fruits.

Two major groups of hormones are:
➤ auxins
➤ gibberellins.

These hormones move through the plant in solution and affect its growth by responding to:
➤ gravity (**gravitropism/geotropism**)
➤ light (**phototropism**).

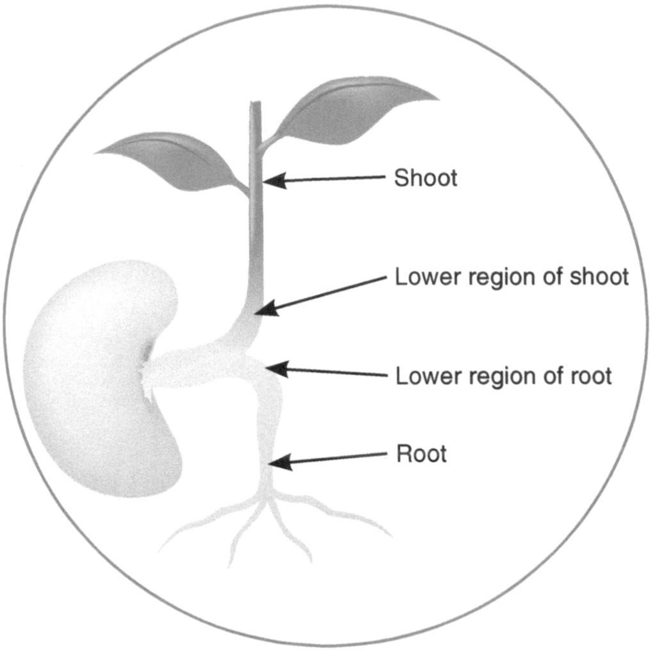

Diagram labels:
- Shoot
- Lower region of shoot
- Lower region of root
- Root

Geotropism

Root

Auxin moves to this side – growth is inhibited so root grows downwards

Response to gravity
➤ **Shoots** grow **against** gravity (**negative gravitropism**).
➤ **Roots** grow **in the direction of gravity** (**positive gravitropism**).

In the roots, auxins inhibit growth in the lower region, which makes the roots grow downwards. The roots anchor the plant in the soil and seek out water and minerals for absorption.

Response to light
➤ **Roots** grow **away** from the light (**negative tropism**).
➤ **Shoots** grow **towards** the light (**positive tropism**).

In the shoot, auxin is made in the tip. When light reaches the tip from all directions, the hormone is equally distributed. When light is **uni-directional** (in this case from the right), the auxin moves to the shaded side, causing the cells here to **elongate**. This makes the shoot bend towards the light.

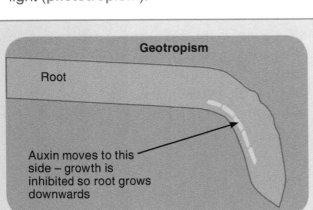

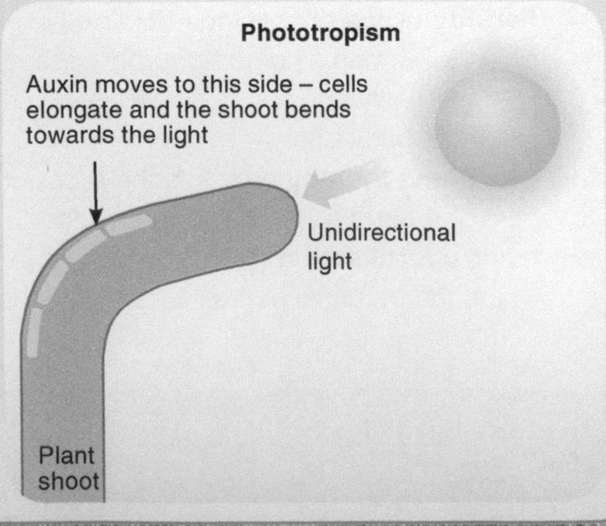

Phototropism

Auxin moves to this side – cells elongate and the shoot bends towards the light

Unidirectional light

Plant shoot

Auxins

➤ **Rooting powder** consists of an auxin that encourages the growth of roots in stem cuttings, so that many new plants can be grown from a parent.

➤ **Tissue culture** technique also requires auxins to promote growth.

➤ **Selective weed killers**, used in agriculture, contain auxins that disrupt the growth patterns of their target plants (which are often broad-leaved rather than narrow-leaved). Therefore the crop plant is not harmed.

Gibberellins

Gibberellins are important in initiating **seed germination**. They can be used to:

➤ end seed dormancy
➤ promote flowering
➤ increase fruit size
➤ produce seedless fruit (parthenocarpy).

Ethene

Ethene is a hydrocarbon that controls cell division. It can be used in the food industry to **control ripening of fruit** while it is being transported or stored.

Keywords

Gravitropism (geotropism) ➤ Growth response in plants against or with the force of gravity

Phototropism ➤ Growth response in plants, towards or away from the direction of incoming light

 One of your required practicals is likely to be investigating the effect of light on the growth of newly germinated shoots.

This is one possible investigation. What other investigations could you carry out?

Experiment to show that shoots grow towards light

1 Cut a hole in the side of a box. Put three cuttings into the box. The cuttings detect light coming from the hole and will grow towards it.

2 Cut a hole in the side of another box. Put three cuttings with foil-covered tips in the box. These shoots can't detect the light so they grow straight up.

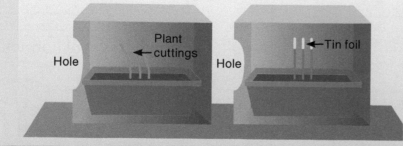

1. What differences would you notice between roots and shoots in the way they respond to increased auxin levels?
HT 2. Which hormones could a farmer apply to her crop to increase the yield?
HT 3. Which hormone could be used to cause plants to flower early?

Mind map

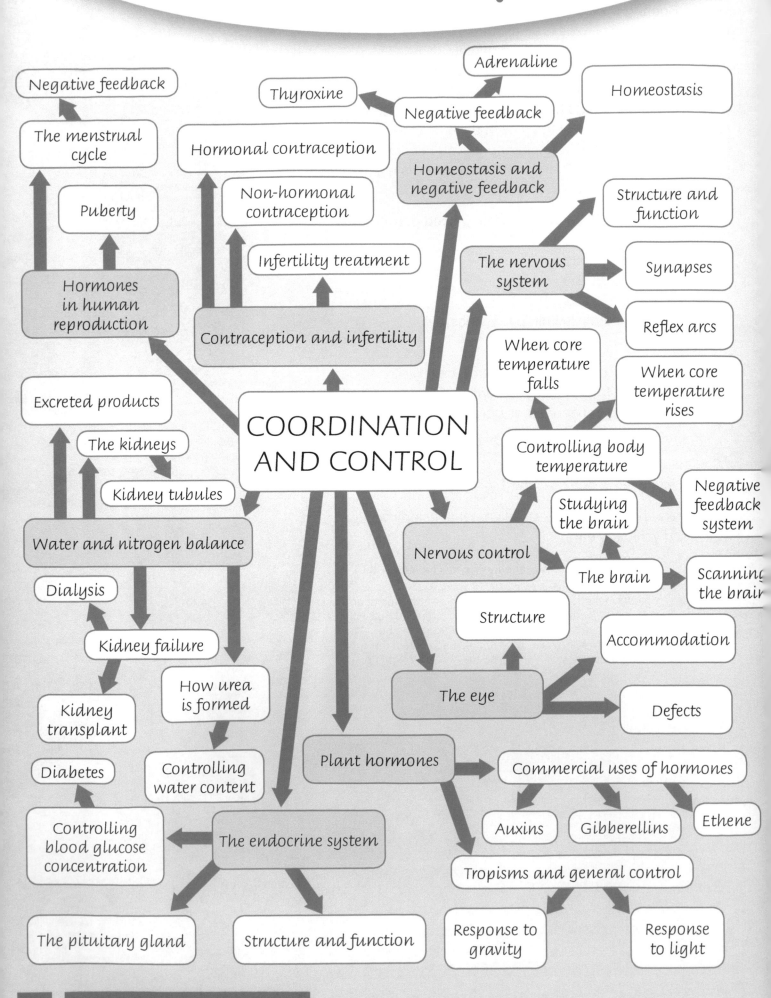

Negative feedback

The menstrual cycle

Thyroxine

Adrenaline

Negative feedback

Homeostasis

Puberty

Hormonal contraception

Homeostasis and negative feedback

Structure and function

Non-hormonal contraception

Hormones in human reproduction

Infertility treatment

The nervous system

Synapses

Reflex arcs

Contraception and infertility

When core temperature falls

When core temperature rises

Excreted products

COORDINATION AND CONTROL

Controlling body temperature

Negative feedback system

The kidneys

Studying the brain

Kidney tubules

Nervous control

The brain

Scanning the brain

Water and nitrogen balance

Dialysis

Structure

Accommodation

Kidney failure

How urea is formed

The eye

Defects

Kidney transplant

Diabetes

Controlling water content

Plant hormones

Commercial uses of hormones

Controlling blood glucose concentration

The endocrine system

Auxins

Gibberellins

Ethene

Tropisms and general control

The pituitary gland

Structure and function

Response to gravity

Response to light

Practice questions

1. The flow chart shows the events that occur during a reflex action.

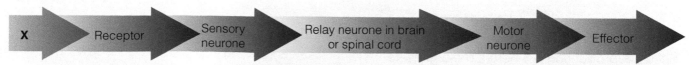

Paul accidentally puts his hand on a pin. Without thinking, he immediately pulls his hand away.

   a) Which component of a reflex arc is represented by the letter X? **(1 mark)**

   b) Give **two** reasons why this can be described as a reflex action. **(2 marks)**

   c) Use the flow chart to describe what happens in this reflex action. **(4 marks)**

2. Look at the graph showing a person's blood sugar levels.

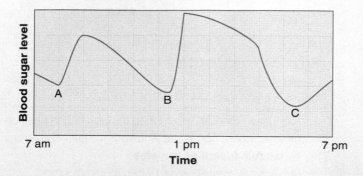

   a) How can we tell from the graph that this person has diabetes? **(2 marks)**

   b) Explain why the person's blood sugar level rises rapidly just after points A and B. **(1 mark)**

   c) Describe what would happen after points A and B if the person did not have diabetes. **(1 mark)**

   d) Explain why the person needed to eat a chocolate at point C. **(1 mark)**

3. Penny is exercising. Many changes are happening in her body. Sweat glands help to control her temperature.

   a) Explain how the sweat glands help to control the temperature of her body. **(2 marks)**

   b) As Penny exercises, she produces carbon dioxide in her cells. What name is given to the process that converts substances into waste products? **(1 mark)**

   c) Penny's liver produces urea. Which substances are changed to form urea? **(1 mark)**

   d) Penny's kidneys process the urea. The route the urea follows is shown below. Write down the missing stages. **(2 marks)**

Sexual and asexual reproduction

One of the basic characteristics of life is **reproduction**. This is the means by which a species continues. If sufficient offspring are not produced, the species becomes **extinct**.

Sexual reproduction

Sexual reproduction is where a male **gamete** (e.g. sperm) meets a female gamete (e.g. egg). This **fusion** of the two gametes is called **fertilisation** and may be *internal* or *external*.

Gametes are produced by **meiosis** in the sex organs (see Module 4).

Asexual reproduction

Sexual reproduction

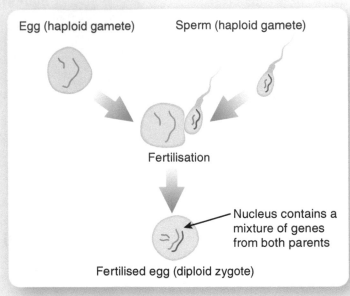

Egg (haploid gamete)   Sperm (haploid gamete)

Fertilisation

Nucleus contains a mixture of genes from both parents

Fertilised egg (diploid zygote)

Asexual reproduction

Asexual reproduction does not require different male and female cells. Instead, genetically identical clones are produced from mitosis. These may just be individual cells, as in the case of yeast, or whole multicellular organisms, e.g. aphids.

Many organisms can reproduce using both methods, depending on the environmental conditions.

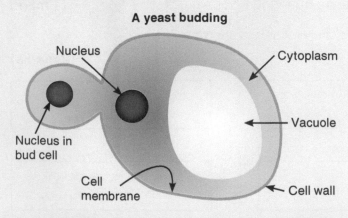

A yeast budding

Nucleus

Cytoplasm

Nucleus in bud cell

Vacuole

Cell membrane

Cell wall

Comparing sexual and asexual reproduction

Advantages of sexual reproduction

> Produces **variation** in offspring through the process of meiosis, where genes are 'shuffled.' Variation is increased by **random fusion of gametes**.
> Survival advantage gained when the environment changes because different genetic types have more chance of producing well-adapted offspring.
> Humans can make use of sexual reproduction through **selective breeding**. This enhances food production.

Disadvantages of sexual reproduction

> Relatively slow process.
> Variation can be a disadvantage in stable environments.
> More resources required than for asexual reproduction, e.g. energy, time.
> Results of selective breeding are unpredictable and might lead to genetic abnormalities from 'in-breeding'.

Advantages of asexual reproduction

> Only one parent required.
> Fewer resources (energy and time) need to be devoted to finding a mate.
> Faster than sexual reproduction – survival advantage of producing many offspring in a short period of time.
> Many identical offspring of a well-adapted individual can be produced to take advantage of favourable conditions.

Disadvantages of asexual reproduction

> Offspring may not be well adapted in a changing environment.

Different reproductive strategies

Malarial parasite

The **plasmodium** is a **protist** that causes malaria. It can reproduce asexually in the human host but sexually in the mosquito.

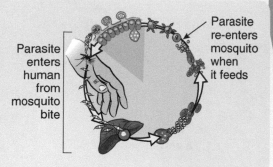

Parasite enters human from mosquito bite

Parasite re-enters mosquito when it feeds

Keywords

Fusion ➤ Joining together; in biology the term is used to describe fertilisation

Internal fertilisation ➤ Gametes join **inside** the body of the female

External fertilisation ➤ Gametes join **outside** the body of the female

Aphid ➤ A type of sap-sucking insect

Strawberry plants and daffodils

These plants can reproduce sexually using flowers, or asexually using **runners** (strawberry) or **bulb division** (daffodils).

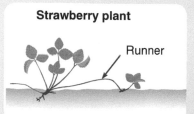

Strawberry plant

Runner

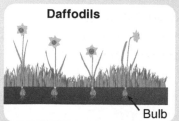

Daffodils

Bulb

Fungi

Many fungi, such as toadstools, mushrooms and moulds, can reproduce sexually (giving variation) or asexually by **spores**.

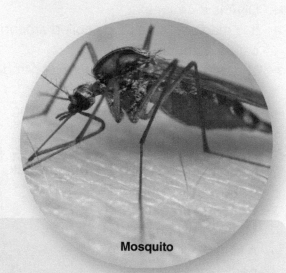

Mosquito

 During your course, you will learn about historical developments in science and technology. It is important to appreciate how understanding develops from previous studies and their publication in scientific journals. Scientists build on each other's work and applications develop from this.

Malaria is a disease that kills many people every year. Treatment via anti-malarial drugs has been available for years and techniques are available to disrupt the parasite's life cycle. At the time of writing, scientists are close to developing a vaccine.

All these developments have come from an understanding of the complex interaction between the plasmodium and mosquito's life cycles. Studies were carried out by many different people and took about ten years to reach their conclusions. Finally, a Scottish physician called Sir Ronald Ross produced evidence for the complete life cycle of the malaria parasite in mosquitoes. For this work, he received the 1902 Nobel Prize in Medicine.

1. Sexual reproduction requires a greater devotion of resources by an organism. So why do so many organisms use it?
2. How do strawberry plants reproduce asexually?
3. State one advantage of asexual reproduction.

DNA

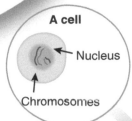

A cell

Nucleus

Chromosomes

DNA and the genome

The nucleus of each cell contains a complete set of genetic instructions called the **genetic code**. The information is carried as genes, which are small sections of DNA found on **chromosomes**. The genetic code controls cell activity and, consequently, characteristics of the whole organism.

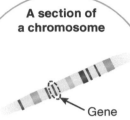

A section of a chromosome

Gene

DNA facts

➤ DNA is a polymer.

➤ It is made of two strands coiled around each other called a **double helix**.

➤ The genetic code is in the form of nitrogenous **bases**.

➤ Bases bond together in pairs forming hydrogen bond cross-links.

➤ The structure of DNA was discovered in 1953 by **James Watson** and **Francis Crick**, using experimental data from **Rosalind Franklin** and **Maurice Wilkins**.

➤ A single gene codes for a particular sequence of **amino acids**, which, in turn, makes up a single **protein**.

DNA double helix

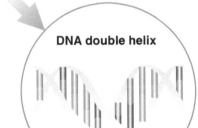

A section of th double helix

Bases

A   T

C   G

G   C

➤ Construct part of a DNA molecule using plasticine. Use different colours to represent bases, sugar and phosphate molecules. Use the model to illustrate how changes in the base sequence can result in new proteins. You could use different colours to represent different amino acids. Make appropriate shapes to represent each component. The diagrams used in this module are a good starting point.

➤ Use your model as a way of learning the names of the different components.

The human genome

The **genome** of an organism is the entire genetic material present in an adult cell of an organism.

The Human Genome Project (HGP)

The HGP was an international study. Its purpose was to map the complete set of genes in the human body.

HGP scientists worked out the code of the human genome in three ways. They:

➤ determined the sequence of all the bases in the genome
➤ drew up maps showing the locations of the genes on chromosomes
➤ produced linkage maps that could be used for tracking inherited traits from generation to generation, e.g. for genetic diseases. This could then lead to targeted treatments for these conditions.

The results of the project, which involved collaboration between UK and US scientists, were published in 2003. Three billion base pairs were determined.

DNA structure and base sequences

The four bases in DNA are A, T, C and G. The code is 'read' on one strand of DNA. Three consecutive bases (a **triplet**) code for one particular amino acid. The sequence of these triplets determines the structure of a whole protein.

The bases are attached to a sugar phosphate **backbone**. These form a basic unit called a nucleotide.

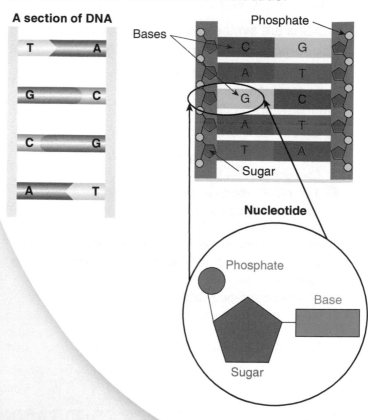

The mapping of the human genome has enabled **anthropologists** to work out historical human migration patterns. This has been achieved by collecting and analysing DNA samples from many people across the globe. The study is called the **Genographic Project**.

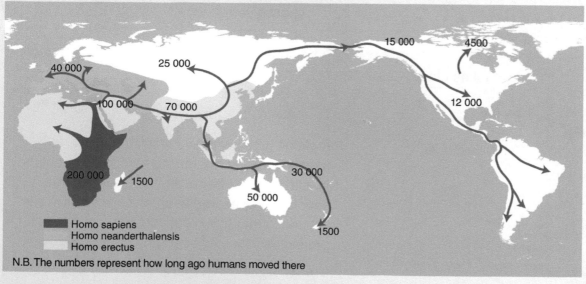

World map showing suggested migration patterns of early hominids

N.B. The numbers represent how long ago humans moved there

1. How has the Human Genome Project advanced medical science?
2. What do genes code for?

The genetic code

DNA structure and base sequences

In **complementary strands** of DNA, T always bonds with A, and C with G.

Proteins are synthesised in structures called **ribosomes**, which are located in the cytoplasm of the cell. In order for the DNA code in the nucleus to be translated as new protein by the ribosomes, the following process occurs.

1. DNA unzips and exposes the bases on each strand.
2. A molecule of **messenger RNA** (**mRNA**) is constructed from one of these template strands.
3. The mRNA, which carries a complementary version of the gene, travels out of the nucleus to the ribosome in cytoplasm.
4. In the ribosome, the mRNA is 'read'. **Transfer RNA** (**tRNA**) molecules carry individual amino acids to add to a growing protein (**polypeptide**).
5. The new polypeptide folds into a unique shape and is released into the cytoplasm.

In the nucleus, **transcription** takes place.

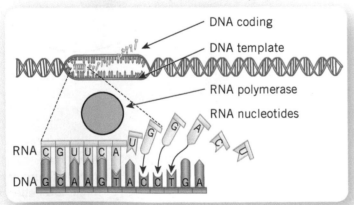

In the ribosome, **translation** takes place.

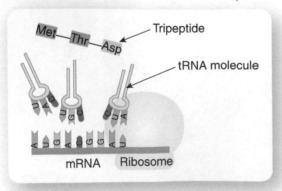

The proteins that are produced carry out a specific function. Examples include enzymes, hormones or structural protein such as **collagen**.

Gene switching

Every cell contains a complete set of genes for the whole body, but only some of these genes are used in any one cell.

> Some genes are not **expressed**. They are said to be 'switched off'. Switching off genes is accomplished through **non-coding** DNA. Most DNA is non-coding.

> The genes that are 'switched on' eventually determine the function of a cell.

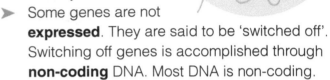
3D view of protein

Mutations

Mutations (genetic variants):

> are changes to the structure of a DNA molecule

> occur continuously during the cell division process or as a result of external influences, e.g. exposure to radioactive materials or emissions such as X-rays or UV light

> usually have a neutral effect as amino acids may still be produced or the proteins produced work in the same way

> may result in harmful conditions or, more rarely, beneficial traits

> result in a change in base sequence and therefore changes in the amino acid sequence and protein structure.

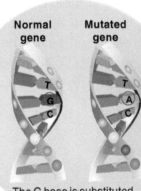
The G base is substituted for an A base

Proteins produced as a result of mutation may no longer be able to carry out their function. This is because they have a different 3D structure. For example, an enzyme's active site may no longer fit with its substrate, or a structural protein may lose its strength.

Changes in the base sequence may be passed on to daughter cells when cell division occurs. This in turn may lead to offspring having genetic conditions.

Monohybrid crosses

Most characteristics or **traits** are the result of multiple alleles interacting but some are controlled by a single gene. Examples include fur colour in mice and the shape of ear lobes in humans. These genes exist as pairs called **alleles** on **homologous chromosomes**.

Alleles are described as being **dominant** or **recessive**.

➤ A **dominant** allele controls the development of a characteristic even if it is present on only one chromosome in a pair.

➤ A **recessive** allele controls the development of a characteristic only if a dominant allele is not present, i.e. if the recessive allele is present on both chromosomes in a pair.

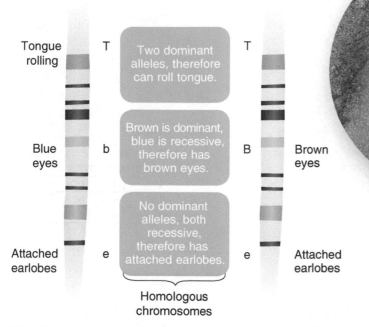

If **both chromosomes** in a pair contain the **same allele** of a gene, the individual is described as being **homozygous** for that gene or condition.

If the chromosomes in a pair contain **different alleles** of a gene, the individual is **heterozygous** for that gene or condition.

The combination of alleles for a particular characteristic is called the **genotype**. For example, the genotype for a homozygous dominant tongue-roller would be **TT**. The fact that this individual is able to roll their tongue is termed their **phenotype**.

Other examples are:

➤ **bb** (genotype), blue eyes (phenotype)
➤ **EE** or **Ee** (genotype), unattached/pendulous ear lobes (phenotype).

When a characteristic is determined by just one pair of alleles, as with eye colour and tongue rolling, it is called **monohybrid inheritance**.

Keywords

HT **Complementary strand** ➤ A sequence of bases that can bond with an equivalent sequence in DNA. Complementary strands can be DNA or mRNA

HT **Polypeptide** ➤ A sequence of peptides, which in turn are smaller chains of amino acids

HT **Transcription** ➤ Process by which a complementary strand of mRNA is made in the nucleus

HT **Translation** ➤ Process occurring in ribosomes where polypeptides are produced from the code in mRNA

Alleles ➤ Alternative forms of a gene on a homologous pair of chromosomes

Homologous chromosomes ➤ A pair of chromosomes carrying alleles that code for the same characteristics

WS During your course, you will be expected to recognise, draw and interpret scientific diagrams.

In this module, the way complementary strands of DNA are arranged can be worked out once you know that base T bonds with A and base C bonds with G.

Can you write out the complementary (DNA) strand to this sequence?

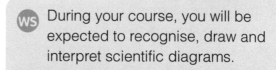

A T T A C G T G A G C C

HT 1. How are the roles of tRNA and mRNA different from each other?
HT 2. Describe the effects of incorrect base sequences on protein manufacture.

Inheritance and genetic disorders

Genetic diagrams

Genetic diagrams are used to show all the possible combinations of alleles and outcomes for a particular gene. They use:

➤ capital letters for dominant alleles
➤ lower-case letters for recessive alleles.

For eye colour, brown is dominant and blue is recessive. So B represents a brown allele and b represents a blue allele.

Example 1

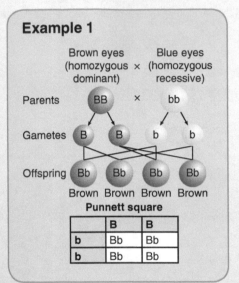

Brown eyes (homozygous dominant) × Blue eyes (homozygous recessive)

Parents: BB × bb

Gametes: B B b b

Offspring: Bb Bb Bb Bb
Brown Brown Brown Brown

Punnett square

	B	B
b	Bb	Bb
b	Bb	Bb

Example 2

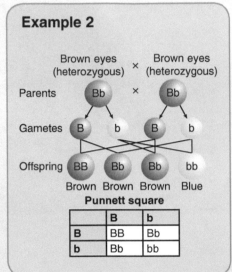

Brown eyes (heterozygous) × Brown eyes (heterozygous)

Parents: Bb × Bb

Gametes: B b B b

Offspring: BB Bb Bb bb
Brown Brown Brown Blue

Punnett square

	B	b
B	BB	Bb
b	Bb	bb

Example 3

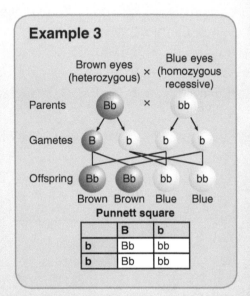

Brown eyes (heterozygous) × Blue eyes (homozygous recessive)

Parents: Bb × bb

Gametes: B b b b

Offspring: Bb Bb bb bb
Brown Brown Blue Blue

Punnett square

	B	b
b	Bb	bb
b	Bb	bb

WS You need to be able to interpret genetic diagrams and work out ratios of offspring.

➤ In Example 2 above, the phenotypes of the offspring are 'brown eyes' and 'blue eyes'. As there are potentially three times as many brown-eyed children as blue-eyed, the phenotypes are said to be in a 3:1 ratio.

➤ In Example 3 above, the ratio would be 1:1 because half of the theoretical offspring are brown-eyed and half blue-eyed. Another way of saying this is that the probability of parents producing a brown-eyed child is 50%, or ½, or 0.5.

Most traits result not from one pair of alleles but from multiple genes interacting, e.g. inheritance of blood groups in the **ABO** system.

HT In exams, you may be asked to construct your own punnett squares to solve genetic cross problems like the ones above.

Family trees

Family trees are another way of showing how genetic traits can be passed on. Here is an example.

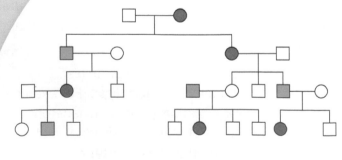

Key:
☐ Male
○ Female
■ Male with trait
● Female with trait

Inheritance of sex

Sex in humans/mammals is determined by whole chromosomes. These are the 23rd pair and are called sex chromosomes. There is an 'X' chromosome and a smaller 'Y' chromosome. The other 22 chromosome pairs carry the remainder of genes coding for the rest of the body's characteristics.

All egg cells carry X chromosomes. Half the sperm carry X chromosomes and half carry Y chromosomes. The sex of an individual depends on whether the egg is fertilised by an X-carrying sperm or a Y-carrying sperm.

If an X sperm fertilises the egg it will become a girl. If a Y sperm fertilises the egg it will become a boy. The chances of these events are equal, which results in approximately equal numbers of male and female offspring.

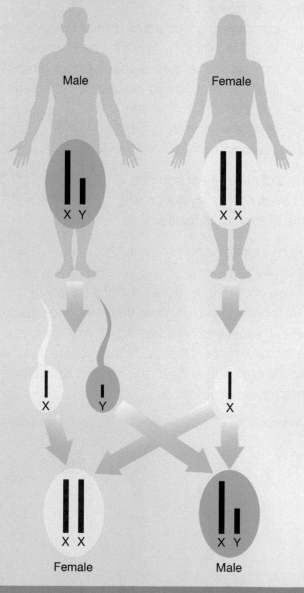

Inherited diseases

Some disorders are caused by a 'faulty' gene, which means they can be **inherited**. One example is **polydactyly**, which is caused by a dominant allele and results in extra fingers or toes. The condition is not life-threatening.

Cystic fibrosis, on the other hand, can limit life expectancy. It causes the mucus in respiratory passages and the gut lining to be very thick, leading to build-up of phlegm and difficulty in producing correct digestive enzymes.

Cystic fibrosis is caused by a recessive allele. This means that an individual will only exhibit symptoms if both recessive alleles are present in the genotype. Those carrying just one allele will not show symptoms, but could potentially pass the condition on to offspring. Such people are called **carriers**.

> Conditions such as cystic fibrosis are mostly caused by **faulty alleles** that are **recessive**.
>
>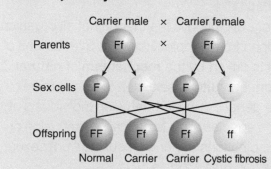
>
> Knowing that there is a 1 in 4 chance that their child might have cystic fibrosis gives parents the opportunity to make decisions about whether to take the risk and have a child. This is a very difficult decision to make.

Technology has advanced and it is now possible to screen embryos for genetic disorders.

➤ If an embryo has a life-threatening condition it could be destroyed, or simply not be implanted if IVF was being applied.

➤ Alternatively, new gene therapy techniques might be able to reverse the effects of the condition, resulting in a healthy baby.

As yet, these possibilities have to gain approval from ethics committees – some people think that this type of 'interference with nature' could have harmful consequences.

1. If a homozygous brown-eyed individual is crossed with a homozygous blue-eyed individual, what is the probability of them producing a blue-eyed child?
2. What is the genotype of a human female?
3. How does the combination of two faulty, cystic fibrosis genes affect the phenotype of the person who possesses them?

Variation and evolution

32

Keywords

Anatomy ➤ The study of structures within the bodies of organisms

Vertebrate ➤ Animal with a backbone

Compress ➤ Squash or squeeze. In geology, this is usually due to Earth movements or laying down sediments

Extinction ➤ When there are no more members of a species left living

Variation

The two major factors that contribute to the appearance and function of an organism are:

➤ **genetic information** passed on from parents to offspring

➤ **environment** – the conditions that affect that organism during its lifetime, e.g. climate, diet, etc.

These two factors account for the large variation we see **within** and **between** species. In most cases, both of these factors play a part.

Evolution

Put simply, evolution is the theory that all organisms have arisen from simpler life forms over billions of years. It is driven by the **mechanism** of **natural selection**. For natural selection to occur, there must be genetic variation between individuals of the same species. This is caused by mutation or new combinations of genes resulting from sexual reproduction (see Module 30).

Most mutations have no effect on the phenotype of an organism. Where they do, and if the environment is changing, this can lead to relatively rapid change.

Evidence for evolution

Evidence for evolution comes from many sources. It includes:

➤ comparing genomes of different organisms

➤ studying embryos and their similarities

➤ looking at changes in species during modern times, e.g. antibiotic resistance in bacteria

➤ comparing the anatomy of different forms

➤ the fossil record.

One of the earliest sources of evidence for evolution was the discovery of fossils.

Further evidence for evolution – the pentadactyl limb

If you compare the forelimbs of a variety of **vertebrates**, you can see that the bone structures are all variations on a five-digit form, whether it be a leg, a wing or a flipper. This suggests that there was a common ancestral form from which these organisms developed.

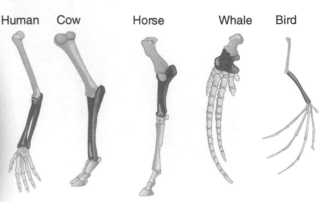

Human   Cow   Horse   Whale   Bird

How fossils are formed

1. When an animal or plant dies, the processes of decay usually cause all the body tissues to break down. In rare circumstances, the organism's body is rapidly **covered** and oxygen is prevented from reaching it. Instead of decay, fossilisation occurs.

2. Over hundreds of thousands of years, further sediments are laid down and **compress** the organism's remains.
3. Parts of the organism, such as bones and teeth, are **replaced by minerals** from solutions in the rock.

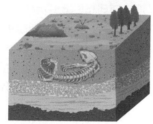

4. Earth upheavals, e.g. **tectonic plate movement**, bring sediments containing the fossils nearer the Earth's surface.

5. Erosion of the rock by wind, rain and rivers exposes the fossil. At this stage, the remains might be found and excavated by **paleontologists**.

Fossils can also be formed from footprints, burrows and traces of tree roots.

By comparing different fossils and where they are found in the rock layers, paleontologists can gain insights into how one form may have developed into another.

Difficulties occur with earlier life forms because many were **soft-bodied** and therefore not as well-preserved as organisms with bones or shells. Any that *are* formed are easily destroyed by Earth movements. As a result of this, scientists cannot be certain about exactly how life began.

Extinction

The fossil record provides evidence that most organisms that once existed have become **extinct**. In fact, there have been at least five **mass extinctions** in geological history where most organisms died out. One of these coincides with the disappearance of dinosaurs.

Causes of extinction include:

➤ **catastrophic events**, e.g. volcanic eruptions, asteroid collisions
➤ changes to the environment over geological time
➤ new **predators**
➤ new **diseases**
➤ new, more successful **competitors**.

Use different-coloured pieces of card to depict one of the ideas covered in this module. Keep the design simple. For example, you could:

➤ show different forms of the pentadactyl limb and how they have evolved from a common ancestor
➤ produce five scenes showing the different stages of fossilisation.

When you have completed your design, explain the process/idea to a revision buddy.

1. State two pieces of evidence that support the theory of evolution through natural selection.
2. Why are fossils so rare?

Darwin and evolution

33

Keywords

Fittest ➤ The most adapted individual or species
Competition ➤ When two individuals or populations seek to exploit a resource, e.g. food. One individual/population will eventually replace the other

Human evolution

Modern man has evolved from a common ancestor that gave rise to all the primates: gorillas, chimpanzees and orangutans. DNA comparisons have shown we are most closely related to the chimpanzee.

The evolution of man can be traced back over the last four to five million years. Over this period of time, the **human form** (hominid) has developed:

➤ a more upright, bipedal stance
➤ less body hair
➤ a smaller and less 'domed' forehead
➤ greater intelligence and use of tools, initially from stone. These tools can be dated using scientific techniques, e.g. radiometric dating and carbon dating.

There have been significant finds of fossils that give clues to human evolution.

1. **Ardi** – at 4.4 million years old, this is the oldest, most complete hominid skeleton.
2. **Lucy** – from 3.2 million years ago, this is one of the first fossils to show an upright walking stance.
3. **Proconsul skull** – discovered by **Mary Leakey** and her husband; the hominid is thought to be from about 1.6 million years ago.

Darwin's theory of evolution through natural selection

Charles Darwin

Within a population of organisms there is a range of variation among individuals. This is caused by genes. Some differences will be beneficial; some will not.

Beneficial characteristics make an organism more likely to survive and pass on their genes to the next generation. This is especially true if the environment is changing. This ability to be successful is called **survival of the fittest**.

Species that are not well adapted to their environment may become extinct. This process of change is summed up in the theory of evolution through **natural selection**, put forward by **Charles Darwin** in the nineteenth century.

Many theories have tried to explain how life might have come about in its present form.

However, Darwin's theory is accepted by most scientists today. This is because it explains a wide range of observations and has been discussed and tested by many scientists.

Darwin's theory can be reduced to five ideas. They are:

➤ variation
➤ survival of the fittest
➤ competition
➤ inheritance
➤ extinction.

This activity should help you memorise Darwin's theory and learn how to apply the features using different scenarios.

➤ Arrange pieces of coloured paper into sets of five. About three lots will do – so fifteen altogether.
➤ On each set, write out the headings for the theory of evolution through natural selection: variation, competition, survival of the fittest, inheritance and extinction.
➤ Now research three case studies relating to natural selection, e.g. warfarin resistance in rats or the shape of shells in Galapagos tortoises. As you read, write down information on each of the five cards.

Darwin's ideas are illustrated in the following two examples.

Example 1: peppered moths

Variation – most peppered moths are pale and speckled. They are easily camouflaged amongst the lichens on silver birch tree bark. There are some rare, dark-coloured varieties (that originally arose from genetic mutation). They are easily seen and eaten by birds.

Competition – in areas with high levels of air pollution, lichens die and the bark becomes discoloured by soot. The lighter peppered moths are now put at a competitive disadvantage.

Survival of the fittest – the dark (melanic) moths are now more likely to avoid detection by predators.

Inheritance – the genes for dark colour are passed on to offspring and gradually become more common in the general population.

Extinction – if the environment remains polluted, the lighter form is more likely to become extinct.

Dark peppered moth

Peppered moth

Example 2: methicillin-resistant bacteria

The resistance of some bacteria to antibiotics is an increasing problem. MRSA bacteria have become more common in hospital wards and are difficult to eradicate.

Variation – bacteria mutate by chance, giving them a resistance to antibiotics.

Competition – the non-resistant bacteria are more likely to be killed by the antibiotic and become less competitive.

Survival of the fittest – the antibiotic-resistant bacteria survive and reproduce more often.

Inheritance – resistant bacteria pass on their genes to a new generation; the gene becomes more common in the general population.

Extinction – non-resistant bacteria are replaced by the newer, resistant strain.

To slow down the rate at which new, resistant strains of bacteria can develop:

➤ doctors are urged not to prescribe antibiotics for obvious viral infections or for mild bacterial infections

➤ patients should complete the full course of antibiotics to ensure that **all** bacteria are destroyed (so that none will survive to mutate into resistant strains).

Species become more and more specialised as they evolve and adapt to their environmental conditions.

The point at which a new species is formed occurs when the original population can no longer interbreed with the newer, 'mutant' population. For this to occur, **isolation** needs to happen.

Speciation

➤ Groups of the same species that are separated from each other by physical boundaries (like mountains or seas) will not be able to breed and share their genes. This is called **geographical isolation**.

➤ Over long periods of time, separate groups may specialise so much that they cannot successfully breed any longer and so two new species are formed – this is **reproductive isolation**.

1. Define evolution and natural selection.
2. State two examples where natural selection has been observed by scientists in recent times.
3. What has to happen to the beneficial genes for a helpful phenotype to spread through a population?

Evolution – the modern synthesis

The pioneers of evolution

Charles Darwin published his findings and theories in a book called *On the Origin of Species* (1859). At the time, the reaction to Darwin's theory, particularly from religious authorities, was hostile. They felt he was saying that 'people were descended from monkeys' (although he wasn't) and that he denied God's role in the creation of man. This meant that Darwin's theory was only slowly and reluctantly accepted by many people in spite of his many eminent supporters. In Darwin's time there wasn't the wealth of evidence we have today from the study of genetics and DNA. (The structure of DNA wasn't made plain until 1953.)

Jean-Baptiste Lamarck had published ideas about the gradual change in organisms before Darwin but he put forward a different approach (now known to be inaccurate). He believed that evolution was driven by the inheritance of acquired characteristics. However, he thought that:

➤ organisms changed during their lifetime as they struggled to survive

➤ these changes were passed on to the offspring.

For example, he said that when giraffes stretched their necks to reach leaves higher up, this extra neck length was passed on to their offspring.

Lamarck's theory was rejected because there was no evidence that the changes occurring in an individual's lifetime could alter their genes and so be passed on to their offspring. The difference between Darwin and Lamarck's approach is shown in the picture.

Lamarck believed that the necks of giraffes stretched during their lifetime to reach food in trees. They then passed this characteristic on to the next generation.

Darwin believed giraffes that had longer necks could reach more food in trees, so they were more likely to survive and reproduce (survival of the fittest).

Alfred Russel Wallace also put forward a theory of evolution by natural selection. In fact, he and Darwin jointly published writings a year before Darwin's *On the Origin of Species*.

Wallace gathered evidence from around the world to back up the theory of natural selection. In particular, he looked at how speciation could occur and wrote extensively about warning colouration in animals.

Gregor Mendel carried out breeding programmes for plants in the middle of the nineteenth century. He discovered that characteristics could be passed down as 'units' (which we now know as genes) in a predictable way.

(HT) (WS) The significance of Mendel's work was only realised after his death. This shows the importance of a scientist's work being published so that others can build on it. It also shows that, sometimes, apparently trivial phenomena and their study can lead to important theories. In this way, our understanding of the world we live in is enhanced.

In your course, you will need to give examples of **how a model can be tested** by **observation** and **experiment**. Mendel's work is a classic example of this because he could show that the ratios of offspring in his pea plant experiments could be accurately predicted.

Other evidence to support the theory of natural selection

Since the nineteenth century, a sequence of findings and techniques has further reinforced our ideas about evolution by natural selection.

➤ In the late nineteenth century, chromosomes were observed during cell division under the microscope.

➤ In the early twentieth century, the link between Mendel's 'units' and genes was made. The idea that genes were found on chromosomes was put forward.

➤ In the mid-twentieth century, the structure of DNA was worked out by Watson and Crick.

Cloning

Studies of genetics have led to many technological advancements. One of these is the ability to **clone** cells, organs or even whole organisms.

Keywords

Clone ➤ A genetically identical cell, tissue, organ or organism

Surrogate mother ➤ An adult female animal that has had an embryo implanted into her uterus. The surrogate gives birth to the offspring but has no genetic relationship to it

Cuttings

When a gardener has a plant with all the desired characteristics, he/she may choose to produce lots of them by taking stem, leaf or root cuttings. The cuttings are grown in a damp atmosphere until roots develop.

Cloning by tissue culture

To mass-produce plants that are genetically identical to the parent plant and to each other, horticulturalists follow this method.

1. Select a parent plant with the characteristics that you want.
2. Scrape off lots of small pieces of tissue into several beakers containing nutrients and hormones.
3. A week or two later there will be lots of genetically identical plantlets growing.
4. Repeat the process.

Embryo transplantation

Embryo transplantation is now commonly used to breed farm animals.

1. Sperm is collected from an adult male with desirable characteristics.
2. A selected female is artificially inseminated with the male animal's sperm.
3. The fertilised egg develops into an embryo that is removed from the mother at an early stage.
4. In the laboratory, the embryo is split to form several clones.
5. Each clone is transplanted into a female who will be the **surrogate mother** to the new calf.

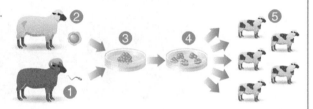

Adult cell cloning

1. The diploid nucleus is taken from a mature cell (ordinary body cell) of the donor organism.
2. The diploid nucleus, containing all of the donor's genetic information, is inserted into an empty egg cell (i.e. an egg cell with the nucleus removed or enucleated). This is nuclear transfer. A tiny electric shock is applied at this stage to trigger cell division in the embryo.
3. The egg cell, containing the diploid nucleus, is stimulated so that it begins to divide by mitosis.
4. The resulting embryo is implanted in the uterus of a 'surrogate mother'.
5. The embryo develops into a foetus and is born as normal.

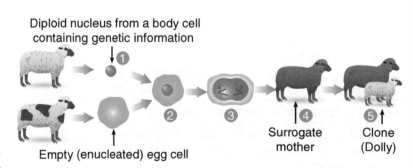

Diploid nucleus from a body cell containing genetic information

Empty (enucleated) egg cell

Surrogate mother

Clone (Dolly)

1. How did Darwin's mechanism of natural selection differ from Lamarck's?
2. How did Mendel's work provide further evidence for Darwin's theory?
3. What is the advantage of cloning plants or animals?

Selective breeding and genetic engineering

Keywords

Herbicide ➤ A chemical applied to crops to kill weeds

HT Vector ➤ Organism, cell or molecule that can be used to transfer DNA from one organism to another

HT Plasmid ➤ A ring of DNA found in bacteria

Selective breeding

Farmers and dog breeders have used the principles of selective breeding for thousands of years by keeping the best animals and plants for breeding.

For example, to breed Dalmatian dogs, the spottiest dogs have been bred through the generations to eventually get Dalmatians. The factor most affected by selective breeding in dogs is probably temperament. Most breeds are either naturally obedient to humans or are trained to be so.

This is the process of selective breeding.

| Select the desired characteristics in parents. | ➤ | Allow the individuals to breed (or cross-pollinate if you are dealing with plants). | ➤ | Select the desired offspring and allow them to become parents of the next generation. |

This process has to be repeated many times to get the desired results.

Advantages of selective breeding
➤ It results in an organism with the 'right' characteristics for a particular function.
➤ In farming and horticulture, it is a more efficient and economically viable process than natural selection.

Disadvantages of selective breeding
➤ Intensive selective breeding reduces the gene pool – the range of alleles in the population decreases so there is **less variation**.
➤ Lower variation reduces a species' ability to respond to environmental change.
➤ It can lead to an accumulation of harmful recessive characteristics (in-breeding), e.g. restriction of respiratory pathways and dislocatable joints in bulldogs.

Examples of selective breeding
Modern food plants

Three of our modern vegetables have come from a single ancestor by selective breeding. (Remember, it can take many, many generations to get the desired results.)

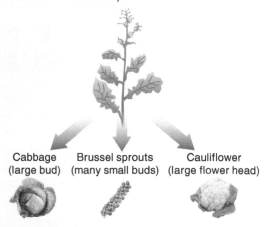

Cabbage (large bud)   Brussel sprouts (many small buds)   Cauliflower (large flower head)

Selective breeding in plants has also been undertaken to produce:
➤ disease resistance in crops
➤ large, unusual flowers in garden plants.

Modern cattle

Selective breeding can contribute to improved yields in cattle. Here are some examples.
➤ **Quantity of milk** – years of selecting and breeding cattle that produce larger than average quantities of milk has produced herds of cows that produce high daily volumes of milk.
➤ **Quality of milk** – as a result of selective breeding, Jersey cows produce milk that is rich and creamy, and can therefore be sold at a higher price.
➤ **Beef production** – the characteristics of the Hereford and Angus varieties have been selected for beef production over the past 200 years or more. They include hardiness, early maturity, high numbers of offspring and the swift, efficient conversion of grass into body mass (meat).

Genetic engineering

All living organisms use the same basic genetic code (DNA). So genes can be transferred from one organism to another in order to deliberately change the recipient's characteristics. This process is called genetic engineering or genetic modification (GM).

Altering the genetic make-up of an organism can be done for many reasons.

➤ **To improve resistance to** herbicides: for example, soya plants are genetically modified by inserting a gene that makes them resistant to a herbicide. When the crop fields are sprayed with the herbicide only the weeds die, leaving the soya plants without competition so they can grow better. Resistance to frost or disease can also be genetically engineered. Bigger yields result.

➤ **To improve the quality of food**: for example, bigger and more tasty fruit.

➤ **To produce a substance you require**: for example, the gene for human insulin can be inserted into bacteria or fungi, to make human insulin on a large scale to treat diabetes.

➤ **Disease resistance**: crop plants receive genes that give them resistance to the bacterium *Bacillus thuringiensis*.

Advantages of genetic engineering

➤ It allows organisms with new features to be produced rapidly.
➤ It can be used to make biochemical processes cheaper and more efficient.
➤ In the future, it may be possible to use genetic engineering to change a person's genes and cure certain disorders, e.g. cystic fibrosis. This is an area of research called gene therapy.

Disadvantages of genetic engineering

➤ Transplanted genes may have unexpected harmful effects on human health.
➤ Some people are worried that GM plants may cross-breed with wild plants and release their new genes into the environment.

HT Producing insulin

The following method is used to produce insulin.

1. The human gene for insulin production is identified and removed using a **restriction enzyme**, which cuts through the DNA strands in precise places.
2. The same restriction enzyme is used to cut open a ring of bacterial **vector** DNA (a **plasmid**).
3. Other enzymes called **ligases** are then used to insert the section of human DNA into the plasmid. The DNA can be 'spliced' in this way because the ends of the strands are 'sticky'.
4. The plasmid is reinserted into a bacterium, which starts to divide rapidly. As it divides, it replicates the plasmid.
5. The bacteria are cultivated on a large scale in fermenters. Each bacterium carries the instructions to make insulin. When the bacteria make the protein, commercial quantities of insulin are produced.

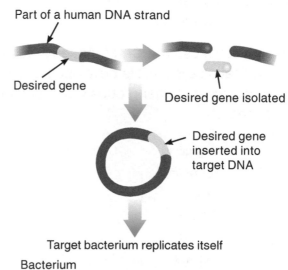

Sometimes, other vectors are used to introduce human DNA into organisms, e.g. viruses. It is important that the hybrid genes are transferred to the host organism at an early stage of its development, e.g. the egg or the embryo stage. As the cells are quite undifferentiated, the desired characteristics from the inserted DNA are more likely to develop.

Design a poster describing the stages of selective breeding.

---

1. What advantages does genetic engineering have over selective breeding?
2. Should we be expanding the range of GM foods we eat? Give one reason **for** this proposal and one reason **against**.

35

There is a huge variety of living organisms. Scientists group or classify them using shared characteristics. This is important because it helps to:

➤ work out how organisms evolved on Earth
➤ understand how organisms coexist in ecological communities
➤ identify and monitor rare organisms that are at risk from extinction.

Classification

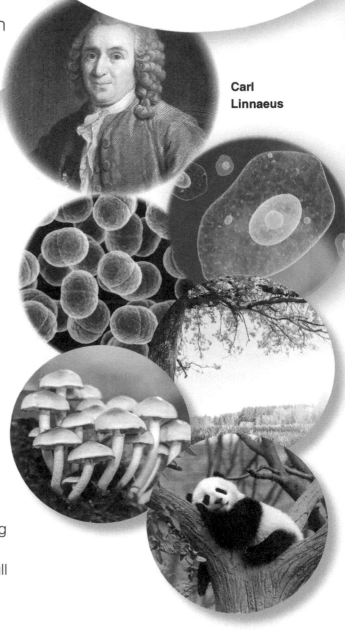

Carl Linnaeus

The origins of classification

In the past, observable characteristics were used to place organisms into categories.

In the eighteenth century, **Carl Linnaeus** produced the first classification system. He developed a hierarchical arrangement where larger groups were subdivided into smaller ones.

Kingdom	Largest group
Phylum	
Class	
Order	
Family	
Genus	
Species	Smallest group

Linnaeus also developed a binomial system for naming organisms according to their genus and species. For example, the common domestic cat is *Felis catus*. Its full classification would be:

➤ Kingdom: *Animalia*
➤ Phylum: *Chordata*
➤ Class: *Mammalia*
➤ Order: *Carnivora*
➤ Family: *Felidae*
➤ Genus: *Felis*
➤ Species: *Catus*.

Linnaeus' system was built on and resulted in a **five-kingdom system**. Developments that contributed to the introduction of this system included improvements in microscopes and a more thorough understanding of the biochemical processes that occur in all living things. For example, the presence of particular chemical pathways in a range of organisms indicated that they probably had a common ancestor and so were more closely related than organisms that didn't share these pathways.

Kingdom	Features	Examples
Plants	Cellulose cell wall Use light energy to produce food	Flowering plants Trees Ferns Mosses
Animals	Multicellular Feed on other organisms	Vertebrates Invertebrates
Fungi	Cell wall of chitin Produce spores	Toadstools Mushrooms Yeasts Moulds
Protoctista Protozoa	Mostly single-celled organisms	Amoeba Paramecium
Prokaryotes	No nucleus	Bacteria Blue–green algae

Keywords

Binomial system ➤ 'Two name' system of naming species using Latin

Common ancestor ➤ Organism that gave rise to two different branches of organisms in an evolutionary tree

Evolutionary trees

Tree diagrams are useful for depicting relationships between similar groups of organisms and determining how they may have developed from common ancestors. Fossil evidence can be invaluable in establishing these relationships.

Here, two species are shown to have evolved from a common ancestor.

The classification diagram below illustrates how different lines of evidence can be used. The classes of vertebrates share a common ancestor and so are quite closely related. Evidence for this lies in comparative anatomy (e.g. the pentadactyl limb) and similarities in biochemical pathways.

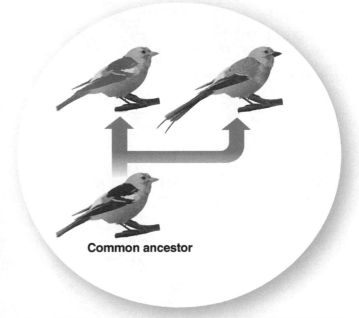

Common ancestor

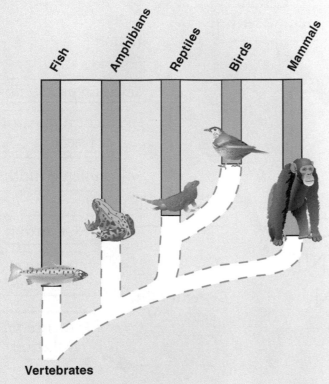

Vertebrates

In more recent times, improvements in science have led to a **three-domain system** developed by **Carl Woese**. In this system organisms are split into:

➤ **archaea** (primitive bacteria)
➤ **bacteria** (true bacteria)
➤ **eukaryota** (including Protista, fungi, plants and animals).

Further improvements in science include chemical analysis and further refinements in comparisons between non-coding sections of DNA.

WS You need to understand how new evidence and data leads to changes in models and theories.

In the case of the three-domain system, a more accurate and cohesive classification structure was proposed as a result of improvements in microscopy and increasing knowledge of organisms' internal structures.

These apes share a common ancestor

1. What do the first and second words in the binomial name of an organism mean?
2. Name four groups of organisms found in the domain *Eukaryota*.
3. It is thought that chimpanzees and humans evolved from a now extinct ape. What term describes this animal?

Mind map

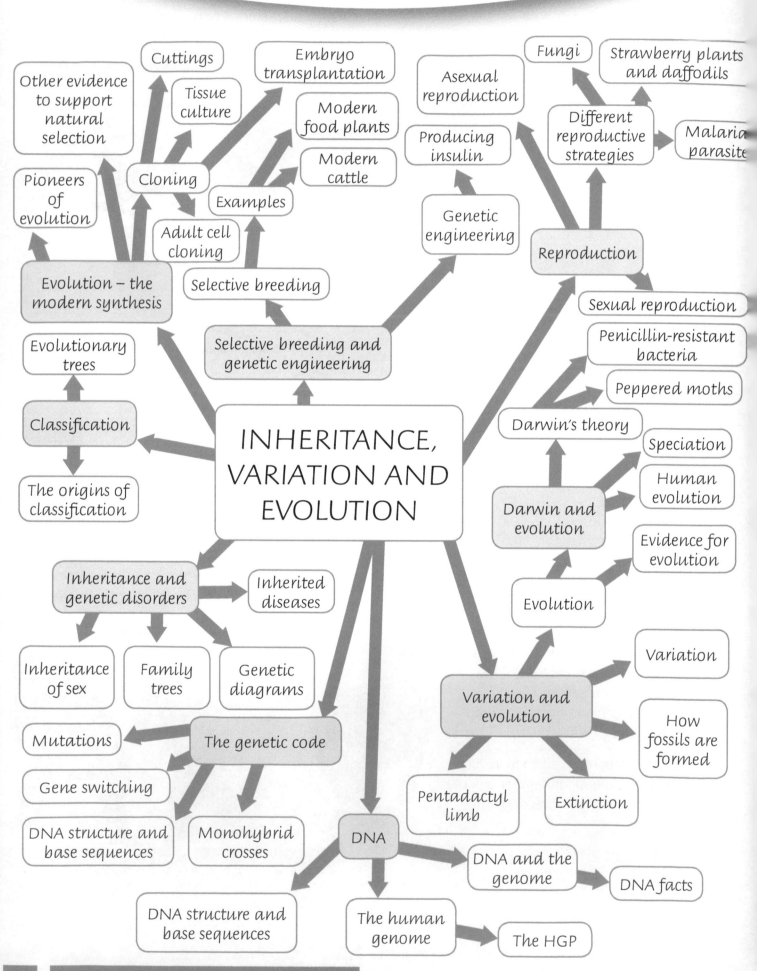

Cuttings

Embryo transplantation

Other evidence to support natural selection

Tissue culture

Fungi

Strawberry plants and daffodils

Asexual reproduction

Modern food plants

Producing insulin

Different reproductive strategies

Malaria parasite

Pioneers of evolution

Cloning

Modern cattle

Examples

Adult cell cloning

Genetic engineering

Reproduction

Evolution – the modern synthesis

Selective breeding

Sexual reproduction

Evolutionary trees

Selective breeding and genetic engineering

Penicillin-resistant bacteria

Peppered moths

Classification

INHERITANCE, VARIATION AND EVOLUTION

Darwin's theory

Speciation

Human evolution

The origins of classification

Darwin and evolution

Evidence for evolution

Evolution

Inheritance and genetic disorders

Inherited diseases

Variation

Inheritance of sex

Family trees

Genetic diagrams

Variation and evolution

How fossils are formed

Mutations

The genetic code

Pentadactyl limb

Extinction

Gene switching

DNA structure and base sequences

Monohybrid crosses

DNA

DNA and the genome

DNA facts

DNA structure and base sequences

The human genome

The HGP

Practice questions

1. Scientists believe that the whale may have evolved from a horse-like ancestor that lived in swampy regions millions of years ago. Suggest how whales could have evolved from a horse-like mammal. In your answer, use Darwin's theory of natural selection. **(4 marks)**

2. The diagram below shows the inheritance of cystic fibrosis in a family.

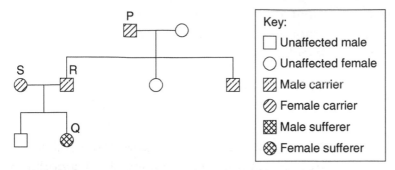

Key:
☐ Unaffected male
◯ Unaffected female
▨ Male carrier
⊘ Female carrier
▨ Male sufferer
⊛ Female sufferer

Cystic fibrosis is caused by a recessive allele, f. The dominant allele of the gene is represented by F.

   a) Give the alleles for person P. **(1 mark)**

   b) Give the alleles for person Q. **(1 mark)**

3. The schematic below shows how insulin can be genetically engineered.

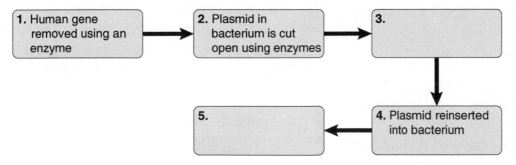

   a) Write the missing steps **3** and **5**. **(2 marks)**

   b) Name the type of enzyme used in steps **1** and **2**. **(1 mark)**

4. The diagram shows a molecule of DNA.

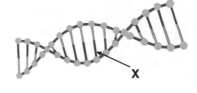

   a) Name the part labelled **X** in the DNA diagram. **(1 mark)**

   b) The 'backbone' of the molecule is arranged in such a way as to make it very stable. Name the term used to describe the shape of DNA. **(1 mark)**

Organisms and ecosystems

Keyword

Environmental resources ➤ Materials or factors that organisms need to survive, e.g. high oxygen concentration, living space or a particular food supply

Communities

Ecosystems are physical environments with a particular set of conditions (**abiotic** factors), plus all the organisms that live in them. The organisms interact through competition and predation. An ecosystem can support itself without any influx of other factors or materials. Its energy source (usually the Sun) is the only external factor.

Other terms help to describe aspects of the environment.

➤ The **habitat** of an animal or plant is the part of the physical environment where it lives. There are many types of habitat, each with particular characteristics, e.g. pond, hedgerow, coral reef.

➤ A **population** is the number of individuals of a species in a defined area.

➤ A **community** is the total number of individuals of all the different populations of organisms that live together in a habitat at any one time.

An organism must be well-suited to its habitat to be able to compete with other species for limited **environmental resources**. Even organisms within the same species may compete in order to survive and breed. Organisms that are specialised in this way are restricted to that type of habitat because their adaptations are unsuitable elsewhere.

Resources that plants compete over include:

➤ light
➤ space
➤ water
➤ minerals.

Animals compete over:

➤ food
➤ mates.
➤ territory.

Interdependence

In communities, each species may depend on other species for food, shelter, pollination and seed dispersal. If one species is removed, it may have knock-on effects for other species.

Stable communities contain species whose numbers fluctuate very little over time. The populations are in balance with the physical factors that exist in that habitat. Stable communities include **tropical rainforests** and ancient oak **woodlands**.

Adaptations

Adaptations:

➤ are special features or behaviours that make an organism particularly well-suited to its environment and better able to compete with other organisms for limited resources

➤ can be thought of as a biological solution to an environmental challenge – evolution provides the solution and makes species fit their environment.

Animals have developed in many different ways to become well adapted to their environment and to help them survive. Adaptations are usually of three types:

➤ **Structural** – for example, skin colouration in chameleons provides camouflage to hide them from predators.

➤ **Functional** – for example, some worms have blood with a high affinity for oxygen; this helps them to survive in anaerobic environments.

➤ **Behavioural** – for example, penguins huddle together to conserve body heat in the Antarctic habitat.

Look at the **polar bear** and its life in a very cold climate. It has:

➤ small ears and large bulk to reduce its surface area to volume ratio and so reduce heat loss
➤ a large amount of insulating fat (blubber)
➤ thick white fur for insulation and camouflage
➤ large feet to spread its weight on snow and ice
➤ fur on the soles of its paws for insulation and grip
➤ powerful legs so it is a good swimmer and runner, which enables it to catch its food
➤ sharp claws and teeth to capture prey.

The **cactus** is well adapted to living in a desert habitat. It:

➤ has a rounded shape, which gives a small surface area to volume ratio and therefore reduces water loss
➤ has a thick waxy cuticle to reduce water loss
➤ stores water in a spongy layer inside its stem to resist drought
➤ has sunken stomata, meaning that air movement is reduced, minimising loss of water vapour through them
➤ has leaves that take the form of spines to reduce water loss and to protect the cactus from predators.

Some organisms have biochemical adaptations. **Extremophiles** can survive extreme environmental conditions. For example:

➤ bacteria living in deep sea vents have optimum temperatures for enzymes that are much higher than 37°C
➤ icefish have antifreeze chemicals in their bodies, which lower the freezing point of body fluids
➤ some organisms can resist high salt concentrations or pressure.

WS During your course you will be asked to suggest explanations for observations made in the field or laboratory. These include:

➤ suggesting factors for which organisms are competing in a certain habitat
➤ giving possible adaptations for organisms in a habitat.

For example, low-lying plants in forest ecosystems often have specific adaptations for maximising light absorption as they are shaded by taller plants. Adaptations might include leaves with a large surface area and higher concentrations of photosynthetic pigments to absorb the correct wavelengths and lower intensities of light.

1. How is an ecosystem different from a habitat?
2. Which is the more stable community – a mixed-leaf woodland or a dry river bed in Africa? What is the reason for this?
3. How is a community different from a population?
4. Give two examples of extremophiles.
5. Why is it important for organisms to be well adapted?

Studying ecosystems 1

Keyword

Sample ➤ A small area or population that is measured to give an indication of numbers within a larger area or population

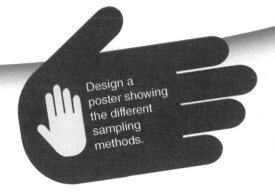

Testing soil pH

Taking measurements in ecosystems

Ecosystems involve the interaction between **non-living** (**abiotic**) and **living** (**biotic**) parts of the environment. So it is important to identify which factors need to be measured in a particular habitat.

Abiotic factors include:
➤ light intensity
➤ temperature
➤ moisture levels
➤ soil pH and mineral content
➤ wind intensity and direction
➤ carbon dioxide levels for plants
➤ oxygen levels for aquatic animals.

Biotic factors include:
➤ availability of food
➤ new predators arriving
➤ new pathogens
➤ one species out-competing another.

Measuring biotic factors – sampling methods

It is usually impossible to count all the species living in a particular area, so a **sample** is taken.

When sampling, make sure you:
➤ **take a big enough sample** to make the estimate good and reliable – the larger the sample, the more accurate the results.
➤ **sample randomly** – the more random the sample, the more likely it is to be representative of the population.

Quadrats

Quadrats are square frames that typically have sides of length 0.5 m. They provide excellent results as long as they are placed randomly. The population of a certain species can then be estimated.

For example, if an average of 4 dandelion plants are found in a 0.25 m² quadrat, a scientist would estimate that 16 dandelion plants would be found in each 1 m² and 16 000 dandelion plants in a 1000 m² field.

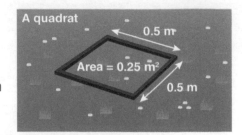

A quadrat

0.5 m

Area = 0.25 m²

0.5 m

Design a poster showing the different sampling methods.

Transects

Sometimes an environmental scientist may want to look at how species change across a habitat, or the boundary between two different habitats – for example, the plants found in a field as you move away from a hedgerow.

This needs a different approach that is systematic rather than random.

1. Lay down a line such as a tape measure. Mark regular intervals on it.
2. Next to the line, lay down a small quadrat. Estimate or count the number of plants of the different species. This can sometimes be done by estimating the percentage cover.
3. Move the quadrat along at regular intervals. Estimate and record the plant populations at each point until the end of the line.

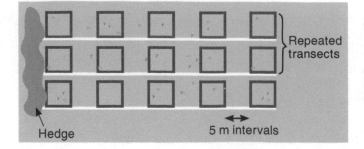

Repeated transects

Hedge

5 m intervals

Sampling methods

Sampling animal populations is more problematic as they are mobile and well adapted to evade capture. Here are four of the main techniques used.

Pooters	Sweepnets	Pitfall traps
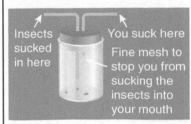		
This is a simple technique in which insects are gathered up easily without harm. With this method, you get to find out which species are actually present, although you have to be systematic about your sampling in order to get representative results and it is difficult to get ideas of numbers.	Sweepnets are used in long grass or moderately dense woodland where there are lots of shrubs. Again, it is difficult to get truly representative samples, particularly in terms of the relative numbers of organisms.	Pitfall traps are set into the ground and used to catch small insects, e.g. beetles. Sometimes a mixture of ethanol or detergent and water is placed in the bottom of the trap to kill the samples, and prevent them from escaping. This method can give an indication of the relative numbers of organisms in a given area if enough traps are used to give a representative sample.

Capture/recapture

The capture/recapture method is sometimes called the **Lincoln index**.

1. Animals are caught humanely – for example, woodlice are caught in traps overnight. Their number is counted and recorded.
2. The animals are marked in some way – for example, water boatmen (a type of insect) can be marked with a drop of waterproof paint on their upper surface.
3. The marked animals are then released back into the population for a suitable amount of time.
4. A second sample is obtained, which will contain some marked animals and some unmarked. The numbers in each group are again counted.

Certain assumptions are made when using capture/recapture data. These include:

➤ no death, immigration or emigration
➤ each sample being collected in exactly the same way without bias
➤ the marks given to the animals not affecting their survival rate, e.g. using paint on invertebrates requires care because if too much is added it can enter their respiratory passages and kill them.

The following formula can then be used to estimate the total population size in the habitat.

$$\frac{\text{population}}{\text{size}} = \frac{\text{number in first sample (all marked)} \times \text{number in second sample (marked and unmarked)}}{\text{number in second sample that were previously marked}}$$

Example

Fifty-six mice are caught in woodland and a small section of fur removed from their tail with clippers. The mice are released. The next evening, a further sample of sixty-two mice are caught. Twenty-five of these mice have shaved tails. What is the total population size?

$$\text{Population size} = \frac{56 \times 62}{25} = 139 \text{ mice}$$

Keys

Correctly identifying species in a sample can be difficult. Using keys like this one can help to identify organisms.

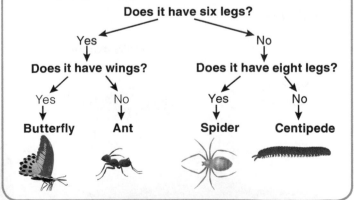

1. In what situation would you use a transect? What information would it give you?

Studying ecosystems 2

Keyword

Indicator species ➤ Organism used as an indicator of pollution

Measuring abiotic factors

Many measurements connected with climate and the weather can be measured using meteorological equipment. Measuring chemical pollutant levels is particularly useful as they give information about concentrations of pollutants. This can be done by **direct chemical tests**.

For example, water can be tested for pH and samples can be assessed for metal ion content, e.g. mercury.

Indicator species

The occurrence of certain **indicator species** can be observed by sampling an area and noting whether the species is present or not, together with the overall numbers of the species. For example, insect larvae and other invertebrates can act as an indicator of water pollution. When sewage works outflow into a stream, this pollutes the water by altering the levels of nitrogen compounds in the stream and reducing oxygen levels. This has an impact on the organisms that can survive in the stream.

Organisms that can cope with pollution include the rat-tailed maggot, the bloodworm, the water louse and the sludge worm. However, some organisms are very sensitive to this type of water pollution so they are not found in areas where the levels are high.

Organisms such as mayfly and stonefly larvae are killed by high levels of water pollution (they cannot tolerate low oxygen levels), so they are indicators of clean water.

Air quality can be measured using lichens as **indicator species**.

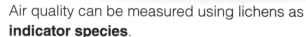

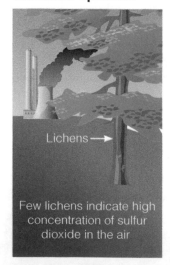

Few lichens indicate high concentration of sulfur dioxide in the air

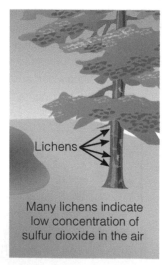

Many lichens indicate low concentration of sulfur dioxide in the air

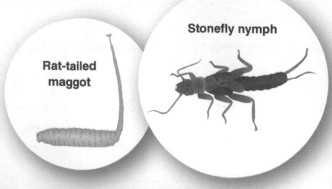

Rat-tailed maggot

Stonefly nymph

Predator–prey relationships

Animals that kill and eat other animals are called **predators** (e.g. foxes, lynx). The animals that are eaten are called **prey** (e.g. rabbits, snowshoe hares).

Many animals can be both predator and prey. For instance, a stoat is a predator when it hunts rabbits and it is the prey when it is hunted by a fox.

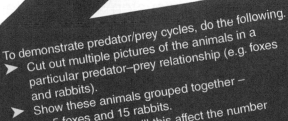

To demonstrate predator/prey cycles, do the following.
- Cut out multiple pictures of the animals in a particular predator–prey relationship (e.g. foxes and rabbits).
- Show these animals grouped together – try 5 foxes and 15 rabbits.
- Add 3 foxes. How will this affect the number of rabbits?
- Explain this to a revision buddy.
- Now show the changes in number of both rabbits and foxes as time goes by. Ask your buddy to rate your explanation.
- Swap roles.

Predator – stoat

Predator – fox

Prey – rabbit

Prey – stoat

In nature there is a delicate balance between the population of a predator (e.g. lynx) and its prey (e.g. snowshoe hare). However, the prey will always outnumber the predators.

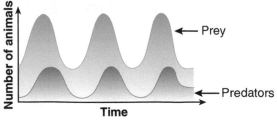

The number of predators and prey follow a classic population cycle. There will always be more hares than lynx and the population peak for the lynx will always come after the population peak for the hare. As the population cycle is cause and effect, they will always be out of phase.

Normal prey population (they outnumber predators)

Predator population increases as plenty of food is available

Decrease in prey population as more are being eaten by increased number of predators

Decrease in predator population as there is now not enough food

39

1. What would the presence of stonefly nymphs tell you about the quality of water in a river?

Materials within ecosystems are constantly being recycled and used to provide the substances that make up future organisms. Two of these substances are water and carbon.

Recycling

The water cycle

Water is a vital part of the **biosphere**. Most organisms consist of over 50% water.

The two key processes that drive the water cycle are **evaporation** and **condensation**.

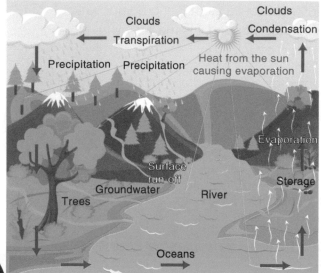

The carbon cycle

The constant recycling of carbon is called the carbon cycle.

➤ Carbon dioxide is removed from the atmosphere by green plants for photosynthesis.

➤ Plants and animals respire, releasing carbon dioxide into the atmosphere.

➤ Animals eat plants and other animals, which incorporates carbon into their bodies. In this way, carbon is passed along food chains and webs.

➤ Microorganisms such as fungi and bacteria feed on dead plants and animals, causing them to decay. The microorganisms respire and release carbon dioxide gas into the air. Mineral ions are returned to the soil through decay.

➤ Some organisms' bodies are turned into fossil fuels over millions of years, trapping the carbon as coal, peat, oil and gas.

➤ When fossil fuels are burned (combustion), the carbon dioxide is returned to the atmosphere.

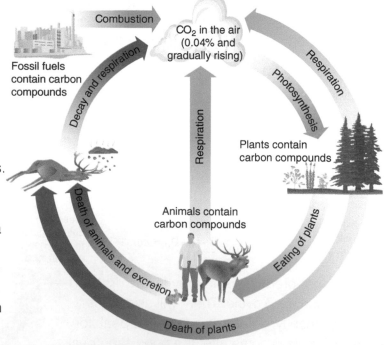

Controlling decomposition

Intensive use of the land removes naturally occurring nitrates, phosphates and other essential mineral nutrients for plants. Farmers can restore these nutrients by applying inorganic fertiliser or using **organic** means, e.g. **compost** or manure. Farmers and gardeners produce compost for the soil from waste plant material.

Keywords

Biosphere ➤ Area on the Earth's crust that is inhabited by living things

Organic ➤ Material obtained from living things; either their dead bodies or their waste. Organic can also refer to a type of farming that avoids overuse of intensive practices such as pesticides and inorganic fertilisers

Compost ➤ Fertiliser produced from the decay of organic plant material

Making compost

In order for decay microorganisms to carry out decomposition, they require:

➤ oxygen (as the process is aerobic)
➤ moisture
➤ a warm temperature.

The presence of any of these factors will increase the rate of decomposition.

One of your required practicals will be to investigate the effect of one factor on the rate of decay.

Biogas digesters

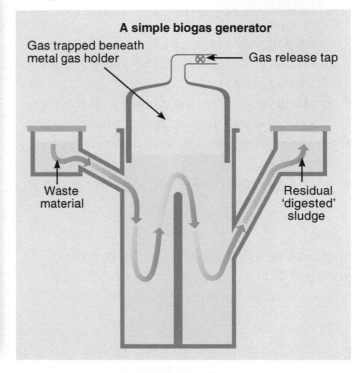

A simple biogas generator

Gas trapped beneath metal gas holder

Gas release tap

Waste material

Residual 'digested' sludge

Anaerobic decay produces methane gas (sometimes called **biogas**). Biogas can be made on a large scale using a continuous-flow method in a digester. Organic material is added daily and the biogas is siphoned off and stored. The remaining solid sludge can be used as fertiliser for crops.

The biogas produced by the digesters can be:
➤ burned to generate electricity
➤ burned to produce hot water and steam central heating systems
➤ used as a fuel for buses.

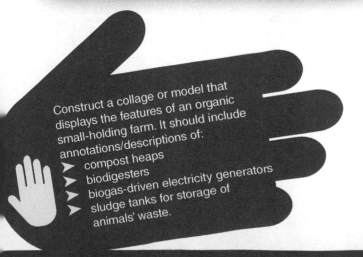

Construct a collage or model that displays the features of an organic small-holding farm. It should include annotations/descriptions of:
➤ compost heaps
➤ biodigesters
➤ biogas-driven electricity generators
➤ sludge tanks for storage of animals' waste.

1. Which two processes drive the water cycle?
2. How can carbon be stored in rocks?
3. Name two processes that add carbon dioxide to the atmosphere.
4. List three conditions needed for the successful decay of organic material.
5. Which combustible chemical is found in biogas?

Environmental change & biodiversity

Environmental change

Waste management

The human population is increasing exponentially (i.e. at a rapidly increasing rate). This is because birth rates exceed death rates by a large margin.

So the use of finite resources like fossil fuels and minerals is accelerating. In addition, waste production is going up:

- **on land**, from domestic waste in landfill, toxic chemical waste, **pesticides** and **herbicides**
- **in water**, from sewage fertiliser and toxic chemicals
- **in the air**, from smoke, carbon dioxide and sulfur dioxide.

Acid rain

When coal or oils are burned, sulfur dioxide is produced. Sulfur dioxide and nitrogen dioxide dissolve in water to produce acid rain.

Acid rain can:

- damage trees, stonework and metals
- make rivers and lakes acidic, which means some organisms can no longer survive.

The acids can be carried a long way away from the factories where they are produced. Acid rain falling in one country could be the result of fossil fuels being burned in another country.

The greenhouse effect and global warming

The diagram explains how global warming can lead to climate change. This in turn leads to lower biodiversity.

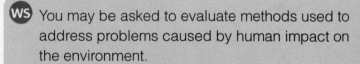

Small amount of infrared radiation transmitted to space

CO_2 and CH_4 in the atmosphere absorb some of the energy and radiate it back to Earth

Rays from the Sun reach Earth and are reflected back towards the atmosphere

The consequences of global warming are:

- a rise in sea levels leading to flooding in low-lying areas and loss of habitat
- the migration of species and changes in their distribution due to more extreme temperature and rainfall patterns; some organisms won't survive being displaced into new habitats, or newly migrated species may outcompete native species. The overall effect is a loss of biodiversity.

WS You may be asked to evaluate methods used to address problems caused by human impact on the environment.

For example, here are some figures relating to quotas and numbers of haddock in the North Sea in two successive years.

	2009	2010
Haddock quota (tonnes)	27 507	23 381
Estimated population (thousands)	102	101

What conclusions could you draw from this data? What additional information would you need to give a more accurate picture?

Create a board game called 'Conservation' that has 100 squares. The object of the game is to improve the quality and quantity of the world's ecosystems. The winner is the first person to reach square 100.

Here is an idea for a **bonus** square.
- Establish a breeding programme for endangered snow-leopards. (Throw the dice again.)

Here is an idea for a penalty square.
- Deforestation of Brazilian rainforest lowers biodiversity. (Go back three spaces.)

Try to grade bonuses and penalties according to their impact.

Biodiversity

Biodiversity is a measure of the number and variety of species within an ecosystem. A healthy ecosystem:

➤ has a large biodiversity
➤ has a large degree of interdependence between species
➤ is stable.

Species depend on each other for food, shelter and keeping the external physical environment maintained. Humans have had a negative impact on biodiversity due to:

➤ pollution killing plants and animals
➤ degrading the environment through deforestation and removing resources such as minerals and fossil fuels
➤ over-exploiting habitats and organisms.

Only recently have humans made efforts to reduce their impact on the environment. It is recognised that maintaining biodiversity is important to ensure the continued survival of the human race.

Impact of land use

As humans increase their economic activity, they use land that would otherwise be inhabited by living organisms. Examples of habitat destruction include:

A marble quarry

➤ farming
➤ quarrying
➤ dumping waste in landfill.

Peat bogs

Peat bogs are important habitats. They support a wide variety of organisms and act as **carbon sinks**.

If peat is burned it releases carbon dioxide into the atmosphere and contributes to global warming. Removing peat for use as compost in gardens takes away the habitat for specialised animals and plants that aren't found in other habitats.

Peat cut and left to dry

Deforestation

Deforestation is a particular problem in tropical regions. Tropical rainforests are removed to:

➤ **release land for cattle and rice fields** – these are needed to feed the world's growing population and for increasingly Western-style diets
➤ **grow crops for biofuel** – the crops are converted to **ethanol-based** fuels for use in petrol and diesel engines. Some specialised engines can run off pure ethanol.

The consequences of deforestation are:

➤ There are fewer plants, particularly trees, to absorb carbon dioxide. This leads to increased carbon dioxide in the atmosphere and accelerated global warming.
➤ Combustion and decay of the wood from deforestation releases more carbon dioxide into the atmosphere.
➤ There is reduced biodiversity as animals lose their habitats and food sources.

Maintaining biodiversity

To prevent further losses in biodiversity and to improve the balance of ecosystems, scientists, the government and environmental organisations can take action.

➤ Scientists establish **breeding programmes** for **endangered species**. These may be captive methods where animals are enclosed, or protection schemes that allow rare species to breed without being poached or killed illegally.
➤ The government sets **limits** on **deforestation** and **greenhouse gas emissions**.

Environmental organisations:

➤ **protect and regenerate** shrinking habitats such as mangrove swamps, heathlands and coral reefs
➤ **conserve and replant** hedgerows around the margins of fields used for crop growth
➤ introduce and encourage **recycling** initiatives that reduce the volume of landfill.

1. Name one pollutant gas that contributes to acid rain.
2. The human population is increasing exponentially – what does this term mean?
3. Why is high biodiversity seen as a good thing?
4. What steps could you take to maintain the biodiversity of a British mixed woodland?

41

Energy and biomass in ecosystems

42

Trophic levels

Communities of organisms are organised in an ecosystem according to their feeding habits.

Food chains show:
➤ the organisms that consume other organisms
➤ the transfer of **energy** and **materials** from organism to organism.

Energy from the Sun enters most food chains when green plants absorb sunlight to **photosynthesise**. Photosynthetic and chemosynthetic organisms are the producers of **biomass** for the Earth. Feeding passes this energy and biomass from one organism to the next along the food chain.

A food chain

Green plant: **producer**

Rabbit: **primary consumer**

Stoat: **secondary consumer**

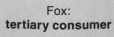

Fox: **tertiary consumer**

The arrow shows the flow of energy and biomass along the food chain.
➤ All food chains start with a **producer**.
➤ The rabbit is a herbivore (plant eater), also known as the **primary consumer**.
➤ The stoat is a carnivore (meat eater), also known as the **secondary consumer**.
➤ The fox is the top carnivore in this food chain, the **tertiary consumer**.

Each consumer or producer occupies a **trophic level** (feeding level).
➤ Level 1 are producers.
➤ Level 2 are primary consumers.
➤ Level 3 are secondary consumers.
➤ Level 4 are tertiary consumers.

Excretory products and uneaten parts of organisms can be the starting points for other food chains, especially those involving **decomposers**.

Decomposers

Pyramids of biomass

Pyramids of biomass deal with the dry mass of living material in the chain. The width of each trophic level can be calculated by multiplying the number of organisms by their dry mass.

Hawk
Thrushes
Slugs
Lettuces

Efficiency of energy transfer

If you know how much energy is stored in the living organisms at each level of a food chain, the efficiency of energy transfer can be calculated.

To do this, divide the amount of energy used usefully (e.g. for growth) by the total amount of energy taken in.

$$\text{energy efficiency (\%)} = \frac{\text{energy used usefully}}{\text{total energy taken in}} \times 100$$

Example: A sheep eats 100 kJ of energy in the form of grass but only 9 kJ becomes new body tissue; the rest is lost as faeces, urine or heat. Calculate the efficiency of energy transfer in the sheep.

$$\text{energy efficiency} = \frac{9}{100} \times 100$$
$$= 9\%$$

Similar efficiency calculations can be performed for biomass.

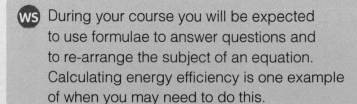 During your course you will be expected to use formulae to answer questions and to re-arrange the subject of an equation. Calculating energy efficiency is one example of when you may need to do this.

For example, you might be given the energy efficiency for a particular trophic level and asked to find out the total energy taken in. In this case:

$$\text{energy efficiency (\%)} = \frac{\text{energy used usefully}}{\text{total energy taken in}} \times 100$$

would be re-arranged to form:

$$\text{total energy taken in} = \frac{\text{energy used usefully}}{\text{energy efficiency}} \times 100$$

Energy

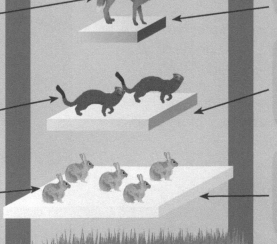

The fox gets the last tiny bit of energy left after all the others have had a share. This explains why food chains rarely have fourth degree or fifth degree consumers – they would not get enough energy to survive.

The stoats run around, mate, excrete, keep warm, etc. They pass on about a tenth of all the energy they get from the rabbits.

The rabbits run around, mate, excrete, keep warm, etc. They pass on about a tenth of all the energy they get from the grass.

The Sun is the energy source for all organisms. However, only about 10% of the Sun's energy is captured in photosynthesis.

Biomass

The fox gets the remaining biomass.

The stoats lose quite a lot of biomass in faeces (egestion), urine (excretion), carbon dioxide and water (respiration).

The rabbits lose quite a lot of biomass in faeces (egestion), urine (excretion), carbon dioxide and water (respiration).

A lot of the biomass remains in the ground as the root system.

Keyword

Decomposers ➤ Microorganisms that break down dead plant and animal material by secreting enzymes into the environment. Small, soluble molecules can then be absorbed back into the microorganism by the process of diffusion

1. What is meant by a trophic level?
2. Why do the trophic levels in a pyramid of biomass decrease in size as you move towards the top carnivore?
3. List three ways in which energy is lost in a food chain.

Farming and food security

Keywords

Yield ➤ The weight of living material harvested in farming and fishing

Eutrophication ➤ A process where nitrates and phosphates enrich waterways, causing massive growth of algae and loss of oxygen

Sustainability ➤ Carrying out human activity, e.g. farming, fishing and extraction of resources from the ground, so that damage to the environment is minimised or removed

43

Farming techniques

Intensive farming is characterised by:
➤ high **yields**
➤ use of mechanised planting and harvesting
➤ extensive use of inorganic fertilisers, pesticides and herbicides.

In order to maximise profits, food production efficiency is the priority. This is achieved by:
➤ restricting energy transfer from animals to the environment, e.g. keeping animals under cover
➤ limiting animals' movement
➤ controlling the temperature of the animals' surroundings.

For example, battery chickens and calves are placed in cages or pens. Fish can be grown in cages and fed high-protein diets.

Many people object to these practices because they think animal welfare has not been given a high enough priority.

Eutrophication

Overusing fertilisers in intensive farming can lead to **eutrophication**.

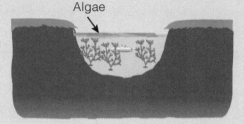

❶ Fertilisers or sewage can run into the water and pollute it. As a result, there are a lot of nitrates and phosphates, which leads to rapid growth of algae.

Algae

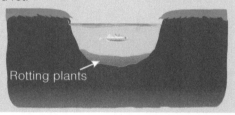

❷ The algal blooms reproduce quickly, then die and rot. They also block off sunlight, which causes underwater plants to die and rot.

Rotting plants

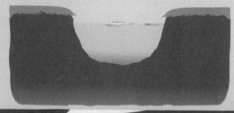

❸ The number of aerobic bacteria increase. As they feed on the dead organisms they use up oxygen. This causes larger organisms and plants to die because they are unable to respire.

Carry out a role-play exercise that mimics a debate between two opposing groups, such as intensive farming versus organic farming or conservationists versus the fishing industry.
➤ Ideally, this activity works best with at least four people (two people representing each group).
➤ Draw up the points you want to make. Include the science you have learned in this topic.
➤ Allow each side to make their case. Then ask an independent 'judge' to make a ruling.
➤ Evaluate the exercise – ask the judge what factors led them to make their decision.

Food security

Food security is a global issue. It recognises the importance of allowing the world's growing population to have access to an adequate diet. Greater industrialisation and higher standards of living are leading to changing diets in countries such as China. So-called 'Western diets' have a higher red meat content. This increases the demand for more land devoted to raising livestock.

Sustainable methods of food production are constantly sought, but many factors need to be taken into account when trying to achieve food security.

➤ Many developed countries have a demand for scarce food resources, meaning that some foods are transported many thousands of miles from their point of origin.
➤ Due to the massive use of pesticides, new pests develop that are resistant to existing chemicals. Pathogenic microorganisms also evolve to become resistant to methods of control such as antibiotics.
➤ Climate change means that some countries are deprived of rainfall, resulting in a failure of harvests. This leads to widespread famine and armed conflicts as populations migrate to find work and food.
➤ Cost of farming methods continues to rise.

In the future, agricultural solutions to the increasing problem of food security might include:

➤ the increased use of hydroponics – a way of growing plants without soil under very controlled conditions
➤ biological control – where natural invertebrate predators are used to control pests
➤ gene technology – including pest-resistant GM crops
➤ fertilisers and pesticides.

Sustainable fishing

Fish stocks are declining across the globe. The problem is so big that unless methods are used to halt the decline, some species (such as the northwest Atlantic cod) might disappear.

Methods to make fishing sustainable include:

➤ governments imposing **quotas** that limit the weight of fish that can be taken from the oceans on a yearly basis – this practice is not always followed and illegal fishing is difficult to combat
➤ increasing the mesh size of nets to allow smaller fish to escape and reach adulthood, so that they can breed.

Biotechnology

Biotechnology, particularly the use of genetic engineering, has allowed new food substances to be produced relatively cheaply. Microorganisms can be grown in large industrial vats.

For instance, the fungus, Fusarium, can be used to produce **mycoprotein**, which is protein-rich and suitable for vegetarians. The fungus is grown on glucose syrup in **aerobic conditions** in a matter of days. Once grown, the biomass is **harvested** and **purified**.

GM crops can increase yields and add nutritional value by increasing the vitamin content of crops such as golden rice.

1. What factors threaten food security globally?
2. What are the advantages and disadvantages of keeping battery hens for food production?
3. Describe how biotechnology has improved food security in modern society.

Mind map

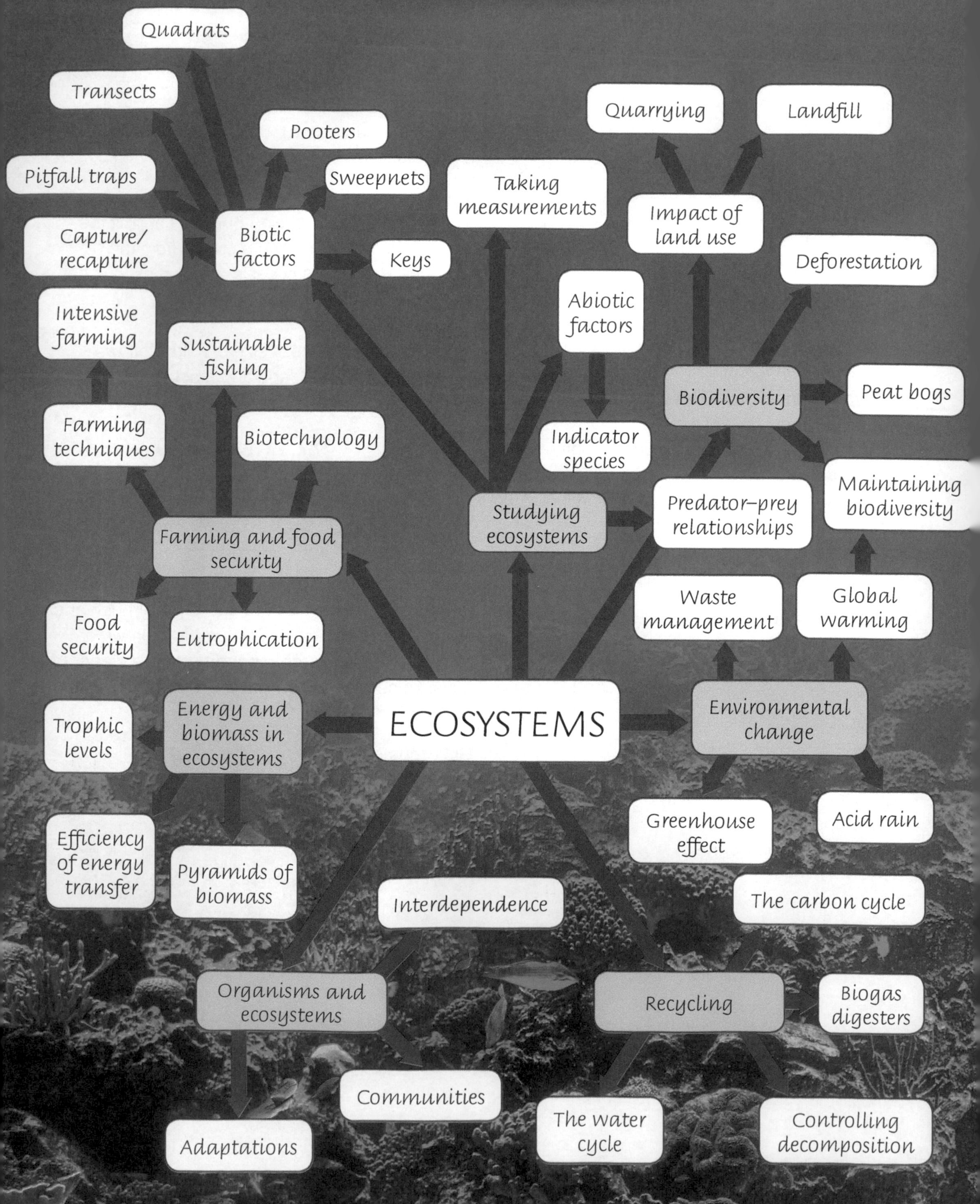

Quadrats

Transects

Pooters

Pitfall traps

Sweepnets

Taking measurements

Quarrying

Landfill

Capture/ recapture

Biotic factors

Keys

Impact of land use

Deforestation

Intensive farming

Sustainable fishing

Abiotic factors

Biodiversity

Peat bogs

Farming techniques

Biotechnology

Indicator species

Maintaining biodiversity

Farming and food security

Studying ecosystems

Predator–prey relationships

Food security

Eutrophication

Waste management

Global warming

Trophic levels

Energy and biomass in ecosystems

ECOSYSTEMS

Environmental change

Efficiency of energy transfer

Pyramids of biomass

Greenhouse effect

Acid rain

Interdependence

The carbon cycle

Organisms and ecosystems

Recycling

Biogas digesters

Communities

The water cycle

Controlling decomposition

Adaptations

Practice questions

1. Carbon is recycled in the environment in a process called the **carbon cycle**. The main processes of the carbon cycle are shown on the right.

   a) Name the process that occurs at stage **3** in the diagram. **(1 mark)**

   b) The UK government is planning to use fewer fossil-fuel-burning power stations in the future. How might this affect the carbon cycle? Use ideas about **combustion** and **fossil fuel formation** in your answer. **(2 marks)**

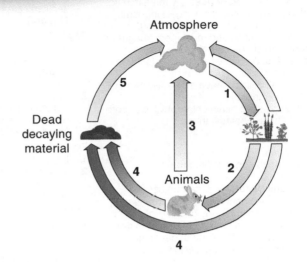

2. An environmental scientist observed and measured a kingfisher and fish population in a county's rivers over 10 years. She recorded her results as a graph.

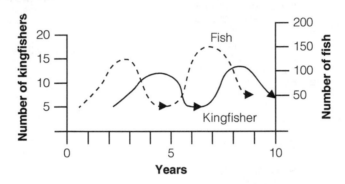

   a) How many fish were recorded in the third year? **(1 mark)**

   b) Describe how the size of the kingfisher population affected the size of the fish population. **(1 mark)**

   HT c) The scientist took her measurements by ringing and observing kingfishers on three rivers in the county. Fish numbers were estimated by counting the different species that anglers landed along the banks of the three rivers. Describe the limitations of these methods and suggest two ways in which the methods could be improved. **(6 marks)**

3. Choose the correct words to complete the following passage about adaptations. **(3 marks)**

environment	population	features	community
characteristics	survival	evolutionary	predatory	suited

   Adaptations are special .... or .... that make a

   living organism particularly well .... to its ....

   Adaptations are part of an .... process that increases a living organism's

   chance of ....

Answers

Page 5
1. **Prokaryotes:** any three bacteria or archaebacteria, e.g. cholera, *E. coli* and salmonella. **Eukaryotes:** any three from the plant, animal, protist or fungal kingdoms, e.g. geranium (plant), tiger (animal), amoeba (protist) and mushroom (fungus).
2. A prokaryote has a DNA loop and plasmids. A eukaryote's DNA is found in the nucleus.
3. True – all cells do have a cell membrane.
4. Mitochondrion.
5. Chloroplasts contain chlorophyll to absorb sunlight for photosynthesis.

Page 7
1. An organ.
2. It is able to contract (shorten).
3. Phloem.
4. **Any one benefit**, e.g. can be used to treat serious conditions, cancer research, organ transplants. Any one objection, e.g. the embryo has the potential to be a living human and shouldn't be experimented with, risk of viral transmission.

Page 9
1. **Any two from:** electron microscopes use electrons to form images, light microscopes use light waves; electron microscopes produce 2D and 3D images, light microscopes use 2D images only; electron microscopes have a magnification up to ×500 000 (2D), light microscopes have a magnification of up to ×1500; electron microscopes are able to observe small organelles, light microscopes are only able to observe cells and larger organelles; electron microscopes enable you to see things at high resolution, light microscopes allow you to see things at low resolution.
2. Food/nutrients, warmth and moisture/water.

Page 11
1. For growth, repair and reproduction.
2. Chromosomes.
3. **Mitosis:** asexual reproduction, repair and growth; **meiosis:** sexual reproduction.
4. DNA/chromosomes and all other organelles.

Page 13
1. **Building** – any one from: converting glucose to starch in plants/glucose to glycogen in animals; synthesis of lipid molecules, formation of amino acids in plants (which are built up into proteins). **Breaking down** – any one from: breaking down excess proteins to form urea, respiration.
2. Aerobic.
3. **Any two from:** transmitting nerve impulses, active transport, muscle contraction/ movement, maintaining a constant body temperature, synthesis of molecules.
4. **Humans:** lactic acid; **yeast:** ethanol and carbon dioxide.
5. To pay back the oxygen debt, i.e. take in oxygen to remove the lactic acid in muscles.

Page 15
1. pH and temperature.
2. A substrate molecule fits exactly into a specific enzyme's active site.
3. It emulsifies fat and helps neutralise acid from the stomach.
4. Amino acids.

Page 17
1. **A:** cell membrane, **B:** DNA, **C:** cell wall, **D:** plasmid. (**1 mark for each correct**)
2. a) Lactic acid increase in muscles causes muscle fatigue/tiredness (**1 mark**); lactic acid is toxic (**1 mark**).
   b) Anaerobic respiration releases lower amounts of energy than aerobic respiration; not enough energy available for extended periods of intense activity. (**1 mark**)

c) Glucose is completely broken down (to carbon dioxide and water). (**1 mark**)
3. a) Denaturing/denaturation. (**1 mark**)
   b) Change in shape of active site (**1 mark**); substrate no longer fits active site (**1 mark**).
4. Answer is 82.61 mm² (**2 marks**) Mean radius = 5.13 mm (**1 mark**); cross-sectional area = 3.14 × (5.13)² = 82.61 mm (**1 mark**)

Page 19
1. A shrew – the proportion of its area to its volume is greater, despite the fact that its **total** body area is less than that of an elephant.
2. The plant cells have a higher water potential/lower solute concentration. The water moves **down** an osmotic gradient from inside the cells into the salt solution.

Page 21
1. Root hair cells.
2. To allow space for gases to be exchanged more freely.
3. So that cells can freely carry water up the plant.

Page 23
1. It would increase the (evapo)transpiration rate.
2. When temperatures are high (during the day) and loss of water through open stomata would cause the plant to dehydrate and wilt.
3. Osmosis.
4. Water enters the xylem at root level due to a 'suction' force caused by evaporation from the leaves, called (evapo) transpiration. The water is drawn up as a column within the stem's xylem.

Page 25
1. Valves prevent backflow of blood and ensure it reaches the heart (especially from parts of the body vertically below the heart).
2. Excess fat (particularly saturated fat) in the diet causes cholesterol to build up and block the coronary artery. This restricts the supply of blood to the heart muscle, which then does not receive enough oxygen and glucose. The heart muscle therefore dies.

Page 27
1. **Red blood cells** do not have a nucleus, are biconcave in shape, contain haemoglobin and carry oxygen. **Lymphocytes** have a nucleus, are irregular in shape and are involved in the immune response.
2. Oxygen is breathed into alveoli, diffuses across alveolar wall and combines with haemoglobin in a red blood cell. Blood enters the heart, from which it is pumped to the respiring muscle tissue. The oxygen diffuses from the red blood cell into a muscle cell, where it reacts with glucose during aerobic respiration.

Page 29
1. Photosynthesis takes **in** carbon dioxide and water, and requires an energy input. It produces glucose and oxygen. Respiration **produces** carbon dioxide and water, and releases energy. It absorbs glucose and oxygen.
2. Balancing investment costs with increased profit from increased yield.

Page 31
1. a, d and e (**1 mark** for each correct answer)
2. a) Healthy man (**1 mark**)
   b) **Any two from:** less oxygen absorbed into blood; longer diffusion distance across alveolar wall; insufficient oxygen delivered to muscles to release energy for exercise (**2 marks**)
3. a) **A:** artery (**1 mark**); **B:** vein (**1 mark**).
   b) **Any one from:** an artery has to withstand/ recoil with higher pressure; elasticity allows smoother blood flow/second boost to blood when recoils. (**1 mark**)
   c) **Any one from:** to prevent backflow of blood; compensate for low blood pressure. (**1 mark**)

Page 33
1. **Any three from:** smoking tobacco, drinking excess alcohol, carcinogens, ionising radiation.
2. The body is more prone to infections.

Page 35
1. Cholera bacteria are found in human faeces, which contaminate water supplies.
2. **Any two reasonable answers, e.g.:** measles, Ebola, athlete's foot.

Page 37
1. Phagocyte ongulfs a pathogen where it is digested by the cell's enzymes.
2. Lymphocytes detect antigen; this triggers production of antibodies; once pathogens are destroyed, memory cells are produced; further infection with same pathogen dealt with swiftly, as antibodies produced rapidly in large numbers. The process can also be triggered via vaccination.

Page 39
1. An antibiotic is a drug that kills bacteria; an antibody is a protein produced by the immune system that kills bacteria.
2. An antiviral alleviates symptoms of an infection; analgesics are painkillers.
3. Do not prescribe antibiotics for viral/ non-serious infections; ensure that the full course of antibiotics is completed.

Page 41
1. Human volunteers; computer simulations/ modelling; cells grown in tissue culture.
2. **Any two from:** cancer treatment, pregnancy testing kits, measuring hormone levels, research.

Page 43
1. **Any three from:** stunted growth, leaf spots, areas of decay, growths/ tumours, malformed stems and leaves, discolouration, presence of pests.
2. Photosynthesis and growth is affected.

Page 45
1. Bacterium – cholera; fungus – athlete's foot; virus – HIV; protist – malaria. (**3 marks** for all answers correct; **2 marks** for 2 or 3 answers correct; **1 mark** for 1 answer correct)
2. a) i) Phagocytes engulf the pathogen, then digest it. (**1 mark**)
      ii) Antibodies lock onto antigen/ pathogen/clump pathogens together. (**1 mark**)
   b) Vaccine contains dead/heat-treated pathogen/microbe. (**1 mark**) Antigen recognised as foreign (**1 mark**); lymphocytes produce antibodies against it (**1 mark**); memory cells remain in system ready to produce antibodies if re-infection occurs (**1 mark**).
   c) Antibiotics don't work against viruses (**1 mark**); over-prescription may lead to antibiotic resistance (**1 mark**).
   d) **Any one from:** antiviral drug/ analgesic (**1 mark**)
3. a) Yes (no marks). Any two for **2 marks**: DDD results in a greater weight loss; 5.8 compared with 3.2; significantly higher than the placebo.
      No (no marks). Any two for **2 marks**: Number of volunteers for the DDD trial is very small compared with the other two trials; more trials need to be carried out; data is unreliable.
   b) In a double blind trial, neither the volunteers nor the doctors know which drug has been given (**1 mark**); this eliminates all bias from the test/scientists cannot influence the volunteers' response in any way (**1 mark**).

Page 47
1. **Any two from:** osmoregulation/water balance; balancing blood sugar levels; maintaining a constant body temperature; controlling metabolic rate.
2. The pituitary.

Page 49

1. Axons/dendrites.
2. They ensure a rapid response to a threatening/harmful stimulus, e.g. picking up a hot plate. As a result, they reduce harm to the human body.
3. Flow diagram with the following labels: (stretch) receptor at knee joint stimulated; sensory neurone sends impulse to spine; intermediate/relay neurone relays impulse to motor neurone; motor neurone sends impulse to (thigh) muscle; (thigh) muscle contracts.

Page 51

1. The frontal lobe, which is part of the cerebral cortex.
2. A CT scan is an X-ray test producing cross-sectional images of the brain using computer programs. A PET scan shows how body tissues are working and what they look like. PET scanners work by detecting radiation given off by radiotracers.
3. Sweating increases, vasodilation, hairs lower.

Page 53

1. The iris.
2. More convex/fatter.
3. An eyeball that is too short/a lens that stays too long and thin.

Page 55

1. It releases a range of hormones that control other processes in the body.
2. In the ovaries and the pituitary (female); in the testes (male).
3. Reduced ability of cells to absorb insulin and therefore high levels of blood glucose. This leads to tiredness, frequent urination, poor circulation, eye problems, etc.

Page 57

1. **Any three from:** glucose, amino acids, fatty acids, glycerol, some water.
2. Makes urine more concentrated.
3. It diffuses down a concentration gradient from the patient's blood into the dialysis fluid.

Page 59

1. FSH acts on the ovaries, causing an egg to mature; oestrogen inhibits further production of FSH, stimulates the release of LH and promotes repair of the uterus wall; LH stimulates release of an egg; progesterone maintains the lining of the uterus after ovulation has occurred and inhibits FSH and LH.
2. Just after menstruation has stopped/from day 7.
3. **Any one from:** production of sperm in testes; development of muscles and penis; deepening of the voice; growth of pubic, facial and body hair.

Page 61

1. Oral contraceptive, hormone injection or implant.
2. They provide a barrier/prevent transferral of the virus during sexual intercourse.
3. Microscopes enable scientists to observe in vitro fertilisation and retrieval of zygote/fertilised egg for implantation in the woman.

Page 63

1. **Shoots:** the auxin promotes growth; **roots:** the auxin inhibits growth.
2. Auxin to kill weeds; gibberellins to promote fruit growth.
3. Gibberellins.

Page 65

1. a) Stimulus. **(1 mark)**
   b) The response is automatic/unconscious **(1 mark)**; response is rapid **(1 mark)**.
   c) Stimulus at receptor triggers sensory neurone to send impulse **(1 mark)**; impulse received in brain and/or spinal cord and signal sent to intermediate/relay neurone **(1 mark)**; impulse then sent to motor neurone **(1 mark)**; motor neurone sends impulse to effector/muscles in hand **(1 mark)**.
2. a) The person's blood sugar level **(1 mark)** fluctuates dramatically **(1 mark)**.
   b) The person has eaten their breakfast at A and their lunch at B. **(1 mark)**
   c) There would be a slight rise in blood sugar level, followed by a swift drop back to normal level. **(1 mark)**
   d) Their blood sugar level had dropped too low/below normal. **(1 mark)**
3. a) They release (more) sweat **(1 mark)**; sweat evaporates, taking heat from the skin **(1 mark)**.
   b) Excretion. **(1 mark)**
   c) Amino acids. **(1 mark)**
   d) **Top box:** Urea enters/is carried in blood **(1 mark)**; **box below it:** Urine stored in bladder **(1 mark)**.

Page 67

1. It allows variation, which gives an evolutionary advantage when the environment changes.
2. Via runners.
3. **Any one from:** produces clones of the parent – if these are successfully adapted individuals, then rapid colonisation and survival can be achieved; only one parent required; fewer resources required than sexual reproduction; faster than sexual reproduction.

Page 69

1. It has allowed the production of linkage maps that can be used for tracking inherited traits from generation to generation. This has led to targeted treatments for these conditions.
2. Proteins.

Page 71

1. mRNA is involved in transcription and carries the coding sequence from the nucleus to the ribosome. tRNA is in ribosomes and involved in the translation process, i.e. building polypeptides/proteins from individual amino acids.
2. Wrong base sequences may lead to incorrect or no protein being produced. This may have consequences for health/wrong base sequence may not affect function of protein produced.

Page 73

1. Zero/0%.
2. XX.
3. Causes production of thick mucus in respiratory pathways and interferes with enzyme production.

Page 75

1. **Any two from:** the fossil record; comparative anatomy (pentadactyl limb); looking at changes in species during modern times; studying embryos and their similarities; comparing genomes of different organisms.
2. The conditions for their formation are rare, e.g. rapid burial and a lower chance of being discovered before they are eroded.

Page 77

1. **Evolution:** the long-term changes seen in species over a long period of time. **Natural selection:** the mechanism by which evolution occurs.
2. **Any two examples**, e.g. Kettlewell's moths; antibiotic resistance in bacteria.
3. They must be passed on by the well-adapted individual that contains them, through reproduction, to the next generation.

Page 79

1. In Darwin's theory, change in the inherited material occurs first. This leads to change in the phenotype/external appearance. Lamarck's theory stated that the physical change occurs first.
2. Mendel showed that characteristics could be passed down as 'units' (genes) in a predictable way.
3. Cloning allows mass production of individuals with beneficial traits in a short space of time.

Page 81

1. Genetic engineering is more precise and it takes less time to see results.
2. **Reason for:** food production improved through increased yields and better nutritional content. **Reason against:** GM plants may spread their genetic material into the wider ecosystem, resulting in, for example, herbicide-resistant weeds/possible harmful effect of GM foods on consumers.

Page 83

1. Genus and species.
2. Invertebrates, vertebrates, protists, higher plants.
3. Common ancestor.

Page 85

1. **Any four from:**
   - adapted to environment/had different characteristics
   - named examples of different characteristics, e.g. some horse-like mammals had more flipper-like limbs (as whales have flippers); idea of competition for limited resources
   - examples of different types of competition; idea of survival of the fittest
   - adaptations being advantageous to living in water/idea that adaptations helped them survive
   - named examples of different adaptations, e.g. some horse-like mammals had more flipper-like limbs that allowed them to swim well in water; idea of inheritance of successful characteristics
   - (named) characteristics/adaptations passed on (through breeding). **(4 marks)**
2. a) Ff
   b) ff
3. a) **Step 3:** section of human DNA inserted into plasmid by ligase enzyme **(1 mark)**. **Step 5:** bacterium replicates/cultivated in fermenter. **(1 mark)**
   b) Restriction enzyme. **(1 mark)**
4. a) Nitrogenous base. **(1 mark)**
   b) Double helix. **(1 mark)**

Page 87

1. A **habitat** is the part of the physical environment in which an animal lives. An **ecosystem** includes the habitat, its communities of animals and plants, together with the physical factors that influence them.
2. **A mixed-leaf woodland:** the biodiversity is greater and so the links between different organisms are more extensive.
3. A **population** is the number of individuals of a species in a defined area. A **community** contains many different populations of species.
4. **Any two examples**, e.g. chemotrophs (live in deep ocean trenches and volcanic vents) and icefish (exist in waters less than 0°C in temperature).
5. So that they survive to reproductive age and pass on their genes to the next generation.

Page 89

1. Where you have a transition from one habitat to the next. It would tell you how numbers of different species vary along the line of the transect.

Page 91

1. The water is relatively clean.

Page 93

1. Condensation and evaporation.
2. As coal.
3. **Any two from:** combustion, animal respiration, plant respiration, microbial respiration (including decay).
4. Moisture, warmth and oxygen.
5. Methane.

Page 95

1. **One from:** sulfur dioxide; nitrogen dioxide.
2. Numbers increase at a rapidly increasing rate.
3. With more species, there are more relationships between different organisms and therefore more resistance to disruption from outside influences.
4. Plant a variety of trees, especially deciduous; protect rare species; make it a Site of Special Scientific Interest (SSSI); protect against invasive species, e.g. red squirrel.

Page 97

1. The level that an organism feeds at, e.g. producer, primary consumer level. It refers to layers in a pyramid of biomass.
2. Most energy is lost to the environment between one trophic level and the next.
3. **Any three from:** respiration; movement; reproduction; maintaining constant body temperature; urine; faeces; not all parts of the animal/plant are eaten.

Page 99

1. Over fishing; demand for rarer foods by developed countries; growing world population; climate change; pesticide resistance.
2. **Advantages:** cheap cost of production; rapid growth of stock. **Disadvantage:** reduced animal welfare.
3. New varieties of plants with greater yields have been produced. Production of mycoprotein uses microorganisms and does not require farming animals.

Page 101

1. a) Respiration. **(1 mark)**
   b) **Any two from:** Fossil fuels represent a carbon 'sink'/they absorbed great quantities of carbon many millions of years ago from the atmosphere; combustion in power stations returns this carbon dioxide; less burning of fossil fuels cuts down on carbon emissions; alternative sources of energy may not return as much carbon dioxide to the atmosphere. **(2 marks)**
2. a) 120–125 **(1 mark)**
   b) When kingfishers increase, fish decrease/when kingfishers decrease, fish increase. **(1 mark)**
   c) This is a model answer, which demonstrates quality written communication (QWC) and would score the full **6 marks:** Three rivers is too small a sample so not all birds would be observed, and ringing birds might affect their survival. Some fish may evade capture by anglers and not all anglers will cooperate with the scientist. Not all species will be caught and kingfishers do not feed on every species of fish. Improvements could be made by sampling many more rivers, using more observers, sampling fish directly, e.g. mark/ recapture technique, observing which species of fish the kingfishers take, or by using more efficient capture methods for fish, e.g. netting.
3. features; characteristics (either way around); suited; environment; evolutionary; survival. **(6 words correct = 3 marks, four or five words correct = 2 marks, two or three words correct = 1 mark, one or 0 words correct = 0 marks.)**

Index